T0191861

Analog Circuits and Signal Processing

Series Editors:

Mohammed Ismail, Dublin, USA
Mohamad Sawan, Montreal, Canada

The Analog Circuits and Signal Processing book series, formerly known as the Kluwer International Series in Engineering and Computer Science, is a high level academic and professional series publishing research on the design and applications of analog integrated circuits and signal processing circuits and systems. Typically per year we publish between 5–15 research monographs, professional books, handbooks, edited volumes and textbooks with worldwide distribution to engineers, researchers, educators, and libraries.

The book series promotes and expedites the dissemination of new research results and tutorial views in the analog field. There is an exciting and large volume of research activity in the field worldwide. Researchers are striving to bridge the gap between classical analog work and recent advances in very large scale integration (VLSI) technologies with improved analog capabilities. Analog VLSI has been recognized as a major technology for future information processing. Analog work is showing signs of dramatic changes with emphasis on interdisciplinary research efforts combining device/circuit/technology issues. Consequently, new design concepts, strategies and design tools are being unveiled.

Topics of interest include:

Analog Interface Circuits and Systems;

Data converters;

Active-RC, switched-capacitor and continuous-time integrated filters;

Mixed analog/digital VLSI;

Simulation and modeling, mixed-mode simulation;

Analog nonlinear and computational circuits and signal processing;

Analog Artificial Neural Networks/Artificial Intelligence;

Current-mode Signal Processing; Computer-Aided Design (CAD) tools;

Analog Design in emerging technologies (Scalable CMOS, BiCMOS, GaAs, heterojunction and floating gate technologies, etc.);

Analog Design for Test;

Integrated sensors and actuators; Analog Design Automation/Knowledge-based Systems; Analog VLSI cell libraries; Analog product development; RF Front ends, Wireless communications and Microwave Circuits;

Analog behavioral modeling, Analog HDL.

More information about this series at http://www.springer.com/series/7381

Kamran Souri • Kofi A.A. Makinwa

Energy-Efficient Smart Temperature Sensors in CMOS Technology

Kamran Souri
SiTime Corp.
Santa Clara, CA, USA

Kofi A.A. Makinwa
Delft University of Technology
Delft, The Netherlands

ISSN 1872-082X   ISSN 2197-1854   (electronic)
Analog Circuits and Signal Processing
ISBN 978-3-319-87286-5   ISBN 978-3-319-62307-8   (eBook)
DOI 10.1007/978-3-319-62307-8

Printed on acid-free paper

This Springer imprint is published by Springer Nature
The registered company is Springer International Publishing AG
The registered company address is: Gewerbestrasse 11, 6330 Cham, Switzerland

Acknowledgments

This thesis is the result of my Ph.D. study at the Electronic Instrumentation Laboratory of Delft University of Technology. In a period of about four and half years, I had the chance to experience a productive and enjoyable time in a friendly and encouraging group. In this page, I would like to dedicate my sincere gratitude to all of those who helped and supported me during the past several years.

I would like to start by thanking my supervisor, Kofi Makinwa, for his continuous encouragement, guidance, and support. In particular, I very much enjoyed our informal brainstorming chats, which resulted in many fruitful ideas and created a clear, solid path forward during my Ph.D. study. Thank you Kofi for trusting me and introducing me to the field of precision analog circuit design.

I am also very grateful to Youngcheol Chae for his friendship and technical advice, and I wish him great success with his academic career. Although I didn't get a chance to work with Michiel Pertijs in person, I would like to take the opportunity to appreciate his work on the precision smart temperature sensors, which formed a solid foundation for my research.

This thesis would not have been possible without the help and support of different people at various branches of NXP Semiconductors. In particular, I must thank Frank Thus (now with Broadcom), Hamid Bonakdar, Anton Tombeur, Paul Noten, Jim Caravella (now with Dialog Semiconductors), Jim Spehar, Brad Gunter, Heimo Scheucher, and Youri Ponomarev (now with Analog Devices).

I wish to thank all my colleagues and friends at the Electronic Instrumentation Laboratory for providing a friendly and pleasant work environment. I thank Joyce, Zu-Yao, Qinwen, Caspar, Junfeng, Sha Xia, Ugur, Burak, Bahman, Zhichao, Saleh, Navid, Mina, and Arvin. My special thanks go to Mahdi Kashmiri for being a great colleague. I truly enjoyed our never-ending coffee-time discussions, and I would never forget our *oven-room* moments during the ISSCC submission deadlines.

I am very grateful to Morteza Alavi for his friendship and unconditional help with following up various defence-related matters while I was in the United States. My particular thanks also go to my dear friend, Sanaz Saeid, for her friendship and support over the past several years.

The burden of writing a Ph.D. thesis becomes unbearable when it is concurrent with relocation and starting a new job. I would like to thank my managers at SiTime, Sassan Tabatabaei and Vinod Menon, for their support and understanding of my situation during this period. I would also like to thank Meisam Roshan for his encouragement. I would also appreciate the help by Saleh Heidary and Vincent van Hoek for proofreading of this thesis.

My sincere thanks go to my family and especially to my parents. I appreciate your support and encouragement throughout these years. I am also very grateful to my in-laws for motivating me towards the end of this journey. I must thank my brother Kianoush for his love and ongoing encouragement over the years. I am also indebted to Darioush Keyvani for his continuous support and advice, and for being the first one to introduce me to the field of integrated circuit design.

Last but not least, I would like to express my deepest gratitude to my wife, Sara, for her unconditional love and support during my study, and in particular during the thesis writing period. This work would have never been finished without your persistent encouragement.

Mountain View, CA, USA   Kamran Souri
May 2017

Contents

1   **Introduction**   1
   1.1   Motivation   1
   1.2   Challenges in Wireless Sensing   5
   1.3   CMOS-Compatible Sensing Elements   6
      1.3.1   Bipolar Junction Transistors (BJTs)   6
      1.3.2   Resistors   8
      1.3.3   Electro-Thermal Filters (ETFs)   9
      1.3.4   MOSFETs   10
      1.3.5   Dynamic Threshold MOSFETs (DTMOSTs)   11
   1.4   Energy Efficiency and Resolution FoM   13
   1.5   Prior-Art and Choice of Sensing Element   14
   1.6   Thesis Organization   15
   References   16

2   **Readout Methods for BJT-Based Temperature Sensors**   19
   2.1   Introduction   19
   2.2   Operating Principle of BJT-Based Sensors   19
      2.2.1   Temperature Characteristics of BJTs   20
   2.3   Generic BJT Readout   23
      2.3.1   Topology   23
      2.3.2   ADC Resolution   25
   2.4   Energy Efficiency of BJT-Based Sensors   26
      2.4.1   Efficiency Limits of a BJT-Based Front-End   26
      2.4.2   Energy Efficiency Gap   31
      2.4.3   ADC Topology   32
   2.5   Conclusions   34
   References   35

3   **Energy-Efficient BJT Readout**   37
   3.1   Introduction   37
   3.2   Proposed Sensor Topology   38
      3.2.1   ADC's Resolution Requirement   40

3.3   The Zoom-ADC: An Energy-Efficient ADC 42
    3.3.1   Introduction 42
    3.3.2   Topology 42
    3.3.3   Coarse Converter.. 42
    3.3.4   Fine Converter 44
    3.3.5   System-Level Considerations 46
        3.3.5.1   Redundancy and Guard-Banding 46
        3.3.5.2   Number of Cycles 48
        3.3.5.3   Signal Swing...................................... 49
        3.3.5.4   Integrator Gain 51
        3.3.5.5   DAC Mismatch 52
3.4   Curve Fitting and Trimming 53
3.5   Conclusions 57
References 58

4   BJT-Based, Energy-Efficient Temperature Sensors 59
4.1   A Micropower Temperature Sensor 59
    4.1.1   Analog Front-End.. 60
        4.1.1.1   Topology 60
        4.1.1.2   Effect of Forward Current Gain β_F 61
        4.1.1.3   Offset Cancellation 62
        4.1.1.4   Opamp Topology 62
        4.1.1.5   Precision Issues.................................... 63
    4.1.2   Zoom ADC... 64
        4.1.2.1   Topology 64
        4.1.2.2   Implementation 64
    4.1.3   Measurement Results 67
4.2   An Energy-Efficient Temperature Sensor 69
    4.2.1   Improving Energy Efficiency................................... 70
    4.2.2   An Energy-Efficient Integration Scheme 71
    4.2.3   Implementation 72
        4.2.3.1   Circuit Diagrams 72
        4.2.3.2   Precision Techniques 73
    4.2.4   Realization and Measurements................................ 75
    4.2.5   Thermal Calibration 76
    4.2.6   Voltage Calibration 76
    4.2.7   Batch-to-Batch Spread and Plastic Packaging................. 78
    4.2.8   Noise and ADC Characteristics 79
    4.2.9   Comparison to Previous Work 80
4.3   Sensing High Temperatures 81
    4.3.1   Analog Front-End.. 82
    4.3.2   ADC Design 84
    4.3.3   Measurement Results 85
4.4   Conclusions 87
References 88

5 All-CMOS Precision Temperature Sensors 91
  5.1 DTMOSTs as Sensing Element 92
    5.1.1 Operating Principle 92
    5.1.2 Temperature Sensor Design 93
    5.1.3 Measurement Results 94
  5.2 A Sub-1V All-CMOS Temperature Sensor 97
    5.2.1 Sensor Front-End 98
    5.2.2 Accuracy Issues 99
    5.2.3 System Diagram 100
    5.2.4 Power Domains 100
    5.2.5 Inverter-Based Zoom ADC 102
    5.2.6 Prototype and Measurement Results 104
  5.3 Conclusions 107
  References 107

6 Conclusions 109
  6.1 Main Findings 109
  6.2 Other Applications of This Work 111
  6.3 Future Work 112
  References 113

Index 115

About the Authors

Kamran Souri was born in Tabriz, Iran, in 1980. He received his B.Sc. in Electronics and M.Sc. in Telecommunication Systems from Amirkabir University of Technology, Iran, in 2001 and 2004, respectively. In Sept. 2007, he joined the Electronic Instrumentation Laboratory (EI-Lab), TU-Delft, where he received his M.Sc. degree (cum laude) in Micro-electronics in 2009 and Ph.D. degree in 2016 for his research on energy-efficient smart temperature sensors in CMOS technology.

From 2001 to 2007, he worked at PSP-Ltd, Tehran, Iran, designing embedded systems for use in high-quality audio/video systems and KVM switches. From 2008 to 2009, he was an intern at NXP Semiconductors, Eindhoven, designing energy-efficient temperature sensors for use in RFID tags. Since 2014, he has been with SiTime Corp., Santa Clara, United States, where he is currently a Principal Circuit Design Engineer, focusing on the design of MEMS-based oscillators.

Dr. Souri was the recipient of the IEEE Solid-State Circuits Society Predoctoral Achievement Award in 2013. He has also served as the technical reviewer for several journals in the field, among them the *IEEE Journal of Solid-State Circuits* (JSSC), *Analog Integrated Circuits and Signal Processing* (AICSP), and the *IEEE Transactions on Circuits and Systems* (TCAS).

Kofi A.A. Makinwa received his B.Sc. and M.Sc. degrees from Obafemi Awolowo University, Nigeria, in 1985 and 1988, respectively. In 1989, he received an M.E.E. degree from the Philips International Institute, the Netherlands, and in 2004, a Ph.D. degree from Delft University of Technology, the Netherlands.

From 1989 to 1999, he was a Research Scientist with Philips Research Laboratories, Eindhoven, the Netherlands, where he worked on interactive displays and digital recording systems. In 1999, he joined Delft University of Technology, where he is currently an Antoni van Leeuwenhoek Professor and Head of the Microelectronics Department. His main research interests are in the design of precision mixed-signal circuits, sigma-delta modulators, smart sensors, and sensor interfaces. This has resulted in 12 books, 25 patents, and over 200 technical papers.

Kofi Makinwa is the Analog Subcommittee Chair of the International Solid-State Circuits Conference (ISSCC). He is also on the program committees of the VLSI Symposium, the European Solid-State Circuits Conference (ESSCIRC), and the Advances in Analog Circuit Design (AACD) workshop. He has been a guest editor of the *Journal of Solid-State Circuits* (JSSC) and a distinguished lecturer of the IEEE Solid-State Circuits Society. For his doctoral research, he was awarded the 2005 Simon Stevin Gezel Award from the Dutch Technology Foundation. He is a co-recipient of 14 best paper awards, from the JSSC, ISSCC, VLSI, and Transducers, among others. At the 60th anniversary of ISSCC he was recognized as a top-10 contributor. He is an IEEE Fellow, an alumnus of the Young Academy of the Royal Netherlands Academy of Arts and Sciences, and an elected member of the IEEE Solid-State Circuits Society AdCom, the society's governing board.

Summary

Nowadays, smart temperature sensors, i.e., sensors with digital outputs, are widely used in various systems. Integrating smart sensors into wireless systems such as RFID tags or wireless sensor networks (WSNs) enables wireless temperature sensing, which in turn opens up a wide range of new applications. This thesis describes the requirements, design, and implementation of smart temperature sensors for use in wireless temperature sensing.

In Chap. 1, an introduction to wireless temperature sensing and its requirements is given. Typically, a wireless node is either powered by a battery or scavenges its energy from the environment, e.g., from an external RF magnetic field. Due to the limited amount of energy available, energy efficiency of the integrated sensor restricts either the battery's lifetime or the operating range of the wireless node. On the other hand, mass production imposes stringent requirements on the cost, which calls for CMOS-compatible sensors. To obtain sufficient accuracy, however, CMOS sensors often require time-consuming (and thus costly) *calibration*: a process in which the sensor's output is compared with that of a reference sensor at a number of known temperatures. The information obtained during calibration is then used to *trim* the sensor, thereby improving its accuracy. A short survey of various CMOS compatible choices is presented. It is shown that substrate PNPs are suitable candidates for wireless temperature sensing. They are power-efficient and exhibit a well-defined process spread, which can be effectively trimmed at a single temperature. However, they require supply voltages greater than ≈ 1.2 V, making them ill suited for low-voltage applications and nano-scale CMOS processes. A promising alternative is to bias a MOSFET in the subthreshold region, while its body and gate terminals are shorted. This so-called DTMOST configuration enables sub-1V operation while exhibiting less spread when compared to the bulk configuration. Finally, to facilitate the comparison between energy efficiency of various temperature sensors, a single figure of merit (FoM) is presented.

In Chap. 2, the operating principle of BJT-based smart temperature sensors is presented. Using the parasitic BJTs available in CMOS, a complementary-to-absolute-temperature (CTAT) voltage V_{BE} and a proportional-to-absolute-temperature (PTAT) voltage ΔV_{BE} can be generated. By properly scaling ΔV_{BE}

(with a scalar α) and combining it with V_{BE}, a reference voltage V_{REF} can then be obtained. In a generic BJT readout, the ratio of $\alpha \cdot \Delta V_{BE}$ and V_{REF} is digitized by means of an analog-to-digital converter (ADC) to generate a PTAT ratio μ. The resolution requirement of the ADC is also discussed. It is shown that almost $\frac{2}{3}$ of the ADS's dynamic range is wasted with this approach. To identify the energy efficiency of existing sensors prior to the start of this research, a study of energy efficiency limits in BJT-based sensors is presented. In this analysis, the ultimate energy efficiency of a BJT-based sensor front-end is calculated and the theoretical limits are defined. Two different approaches based on bias-current and emitter-area scaling are considered. Based on this analysis, a significant energy-efficiency gap, over four orders of magnitude, is observed between the prior-art and theoretical limits. The study of various sensor architectures reveals that, in fact, the reason behind this gap lies in the employed readout circuits, which mostly include $\Delta\Sigma$- or SAR-ADCs. They either suffer from long conversion times and poor power efficiency, or are not capable of providing the target resolution or accuracy. To bridge this efficiency gap, a new readout architecture is clearly required.

In Chap. 3, different BJT-based sensor architectures based on digitizing nonlinear ratios between ΔV_{BE} and V_{BE} (or their combinations) are explored. The required linearization to calculate the PTAT ratio μ is then performed in the digital back-end. Since the coefficient α is digitally implemented, it can also be used for trimming. The employed ADC architectures in these examples, however, often result in more waste of dynamic range than in the generic approach, exacerbating the lack of energy efficiency. To address this issue, a new readout topology based on digitizing the ratio $X = V_{BE}/\Delta V_{BE}$ is proposed. Since temperature changes are rather slow, the ratio X is accurately digitized by a two-step *zoom*-ADC. As X is typically greater than one, it can be expressed as $X = n + \mu'$, where n and μ' correspond to the integer and fractional parts, respectively. First, a full-range SAR conversion obtains the integer n by performing a binary search algorithm, comparing V_{BE} to integer multiples of ΔV_{BE}. This is then followed by a low-range fine $\Delta\Sigma$ converter, whose references are set to n and $n + 1$. In this manner, the ratio μ' can be accurately digitized with high resolution. In contrast to the conventional $\Delta\Sigma$-ADCs, the full-scale range of the fine converter in the zoom-ADC is considerably reduced, which notably relaxes various key requirements such as the number of $\Delta\Sigma$-cycles and the DC gain and swing of the loop filter. In this architecture, both conversion time and power efficiency can be improved, which results in a substantial improvement in energy efficiency. The fact that dynamic correction techniques can be used in the fine conversion phase ensures that the accuracy of the zoom-ADC can be as good as that of conventional $\Delta\Sigma$-ADC architectures.

In Chap. 4, a low-power BJT-based sensor prototype based on a 1st-order switched-capacitor (SC) zoom-ADC is presented. It achieves a resolution of 15 mK in a conversion time of 100 ms while dissipating only 4.6 µA. After a single α-trim at 25 °C, the sensor obtains an inaccuracy of ±0.2 °C (3σ) from −30 to 125 °C. This result shows 11× energy efficiency improvement when compared to sensors with similar accuracy, back in 2011. However, its fine conversion step employs a slow, 1st-order $\Delta\Sigma$ modulator, limiting its energy efficiency. Moreover, each

$\Delta\Sigma$ cycle requires two full clock periods, since V_{BE} and ΔV_{BE} are separately sampled/integrated. To further improve the sensor's energy efficiency, a second prototype is realized which achieves similar resolution in about 16× less conversion time, while drawing 25% less supply current. This is achieved by using a 2nd-order zoom-ADC, combined with a new charge-balancing scheme, whose operation is based on *simultaneous* sampling of V_{BE} and ΔV_{BE}. This allows the use of low-swing, low-power amplifiers. The sensor's energy efficiency is therefore improved by over 20× compared to the first prototype. Using a thermal calibration and digital PTAT trimming at 30 °C, the sensor achieves an inaccuracy of ±0.15 °C (3σ) from −55 to 125 °C. Moreover, a voltage calibration technique based on electrical measurements is also explored, which is significantly faster (only requires 200 ms), while achieving comparable accuracy. The impact of batch-to-batch spread and plastic packaging on sensor's accuracy is investigated as well. As observed, both of them can cause temperature reading shifts in the order of 0.4–0.5 °C from −55 to 125 °C.

In the last part of Chap. 4, a BJT-based sensor prototype for sensing high temperatures (>150 °C) is also demonstrated. It is shown that by optimizing the emitter area and bias current of a substrate PNP, the impact of saturation current I_S at high temperatures can be mitigated. Furthermore, robust circuit techniques are employed to cope with the various leakage currents at such temperatures, which would otherwise impact the accuracy of V_{BE} and ΔV_{BE}, and thus the sensor output. It achieves an inaccuracy of ±0.4 °C (3σ) from −55 to 200 °C, which is similar to that of state-of-the-art sensors capable of operating over such temperature ranges. However, it draws only 22 µA, which is more than an order of magnitude less.

In Chap. 5, the use of DTMOSTs as temperature sensing elements is demonstrated. When operated in weak inversion, the gate-source voltage V_{GS} of a DTMOST is almost half of the base-emitter voltage V_{BE}, thus enabling sub-1V operations. Moreover, compared to a diode-connected MOSFET, the V_{GS}–I_D characteristic of a diode-connected DTMOST is less sensitive to the spread in threshold voltage V_T, making it a promising candidate for realizing accurate temperature sensors. Two sensor prototypes based on such sensing elements are demonstrated in a chosen 160 nm CMOS process. After a single-temperature trim, the first prototype achieves an inaccuracy of ±0.4 °C (3σ) from −55 to 125 °C, and enables an apples-to-apples comparison with BJTs, proving that DTMOSTs are indeed only a factor 2× less accurate. In the second prototype, the low-voltage capability of DTMOSTs is then exploited to realize a sub-1V, sub-µW precision sensor. Employing fully inverter-based SC integrators, a 2nd-order zoom-ADC is realized in the second prototype. It can operate at supply voltages as low as 0.85 V, while drawing only 700 nA. It also maintains the same inaccuracy of ±0.4 °C (3σ) from −40 to 125 °C, after a single-temperature trim. These results prove that DTMOSTs could be considered as the temperature sensors of choice when sub-1V, high accuracy, and energy efficiency are key requirements.

In Chap. 6, the main findings of this work are summarized. These include the development of the zoom-ADC and its application in energy-efficient smart temperature sensors. The final prototype BJT-based sensor achieved a resolution

FoM of 11 pJ $^\circ$C^2 and improved state of the art by a factor of 15$\times$ (in 2012). Another key finding was the fact that DTMOST sensors enable low-voltage operations while being only 2$\times$ less accurate than BJT-based sensors. The final prototype achieved a fairly good energy efficiency, evidenced by a FoM of 14 pJ $^\circ$C^2. The chapter also contains some suggestions for future work: to further improve the energy efficiency, continuous-time (CT) readouts could be considered as promising alternatives to the switched-capacitor circuits. Furthermore, to reduce the cost of over-temperature characterizations, a combination of voltage calibration with integrated heaters could be used to quickly extract the global calibration parameters. Another alternative could be to exploit the high accuracy of thermal-diffusivity (TD) sensors as on-die references during the calibration process. The chapter ends with a discussion of the potential use of the zoom-ADC technique to realize general-purpose ADCs with high energy efficiency.

Chapter 1
Introduction

Temperature is the most often-measured environmental quantity [1]. This is because nearly all physical, chemical, mechanical, and biological systems exhibit some sort of temperature dependence. Temperature measurement and control are therefore critical tasks in many applications. Traditionally, temperature sensors have been implemented with discrete components such as resistance temperature detectors (RTDs), thermistors, or thermocouples. In the last three decades, integrated temperature sensors, particularly in CMOS technology, have become a promising alternative. A sustained research effort has been devoted to the development of compact, low-cost temperature sensors with co-integrated readout circuitry, thus providing temperature information in a digital format. Such *smart* temperature sensors (see Fig. 1.1) are conventional products nowadays [3–7].

There are several advantages associated with smart sensors; firstly, since a digital output is almost mandatory in modern systems, no external analog-to-digital converter (ADC) is required. This higher level of integration reduces component count, and therefore size and, typically, cost. Secondly, in contrast to digital signals, analog signals are prone to interference and thus are not well suited for accurately transmitting data to other blocks in a system. Lastly, by integrating the readout circuit and the sensor on the same chip, on-chip digital post-processing becomes possible, which usually results in simpler systems.

1.1 Motivation

Smart temperature sensors have been around for many years. However, with the recent development of low-power radio systems, wireless temperature sensing has become very attractive, as it opens up a wide variety of new applications. One can think of applications in cold supply chains, monitoring of perishable goods, animal husbandry and agriculture, automotive, building automation, and healthcare.

© Springer International Publishing AG 2018   1
K. Souri, K.A.A. Makinwa, *Energy-Efficient Smart Temperature Sensors
in CMOS Technology*, Analog Circuits and Signal Processing,
DOI 10.1007/978-3-319-62307-8_1

Fig. 1.1 Block diagram of an integrated smart temperature sensor [2]

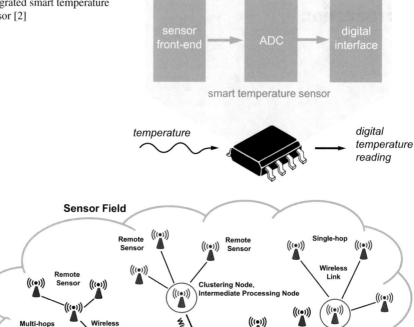

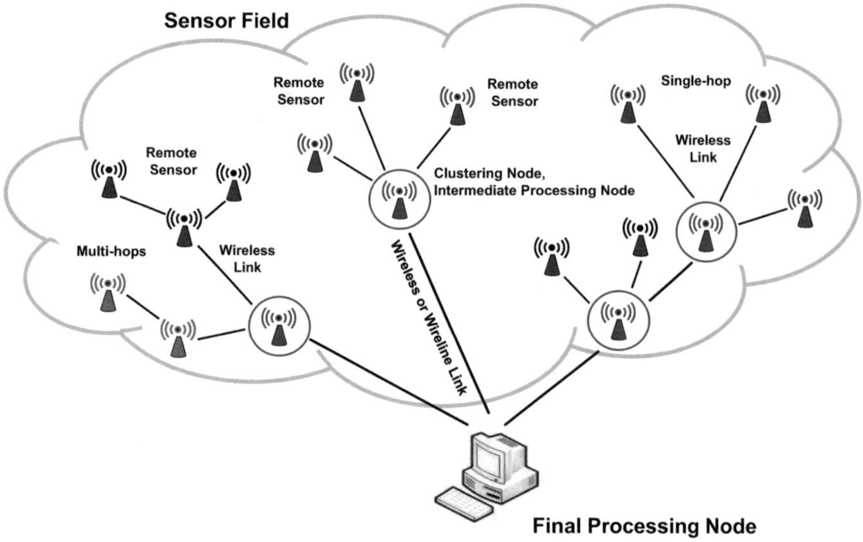

Fig. 1.2 A typical wireless sensor network (WSN) arrangement [8]

Wireless sensor networks (WSNs), which consist of spatially distributed sensor nodes with a wireless communication infrastructure, were introduced in the 2000s [8]. Various physical or environmental quantities such as temperature, sound, humidity, motion, and pressure can be sensed and digitized by the sensor nodes. The digitized signals are then passed through the communication network towards a centralized or distributed control unit for further processing, as shown in Fig. 1.2. As the name WSN suggests, and mainly due to cost reasons and ease of integration, wireless operation is a key feature, which at the same time makes powering the sensor nodes a challenging task. Most WSNs have used battery-powered sensors nodes, while quite recently, nodes based on energy harvesting or scavenging have also been introduced [9].

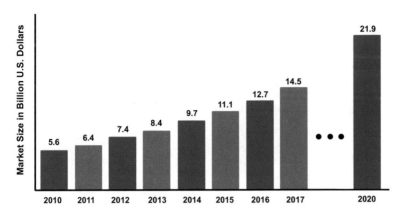

Fig. 1.3 Projected size of the global market for RFID tags from 2010 to 2020 (in billion U.S. dollars) [10]

Another opportunity for wireless sensing has recently emerged through the introduction of radio frequency identification (RFID) technology as a versatile wireless communication platform. RFID has been around for years now and has become a billion dollar market over the last few years and it is still growing. With an estimated $5.6 billion market in 2010, and an average 15% year-on-year growth rate (see Fig. 1.3), the forecasted market in 2020 will exceed $21.9 billion [10]. This shows that RFID technology has achieved solid penetration throughout worldwide commerce, boosted by dynamic growth in the retail apparel sector. The freedom provided by small size and easy positioning, non-line-of-sight wireless operation and powering, and extended read ranges are key features that have made RFID technology so promising.

Apart from its primary application in identification and tracking, RFID has become a pragmatic building block for the internet of things (IoT), thus creating a flood of new applications in numerous industries [11]. According to an IC Market Drivers report in 2016 [12], 30.0 billion Internet connections are expected to be in place worldwide in 2020, with 85% of them being to web-enabled *things*, meaning a wide range of commercial, industrial, and consumer systems, distributed sensors, vehicles, and other connected objects. As reported, IoT applications will fuel strong sales growth in optoelectronics, sensors/actuators, and discrete semiconductors, which are projected to rise by a compound annual growth rate (CAGR) of 26.0% between 2015 and 2019, thus offering a forecasted market of $11.6 billion in 2019.

Most RFID tags consist of two main parts (see Fig. 1.4). The first part is an integrated circuit (IC) to implement the target functionality, e.g., the storing and processing of information, as well as the RF transceiver. This part usually occupies only a small portion of the total area of the tag. The second part, which takes up the bulk of the area, is the antenna, which is required for receiving and transmitting the RF signal. Depending on their source of energy, RFID tags can be classified into *passive* and *active* tags. Active RFID tags include a battery to power the IC, which

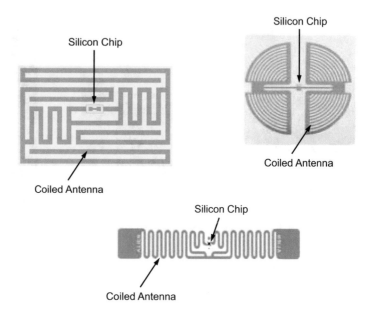

Fig. 1.4 Various samples of RFID tags; each tag is composed of a large antenna and a silicon integrated circuit (IC)

makes autonomous operation possible. In consequence, low power designs, along with brief operating periods, are desirable in order to maximize battery lifetime. Passive RFID tags, in contrast, are not equipped with a battery and consequently, autonomous operation is not possible. Instead, the power required to operate the tag is scavenged from an external magnetic/electromagnetic field, transmitted by a reader. The energy absorbed via an antenna from the field used to power the tag, thus, enabling data transmission and other functionalities. In other words, the antenna of a passive RFID tag is used to transfer information as well as to receive power.

The choice of RFID tag type depends on the target application. Battery-equipped or active RFID tags can communicate over long distances, up to 100 m or more. Furthermore, they can operate continuously. However, they have limited lifetime (typically 1–4 years), significantly higher production costs, e.g., few dollars and larger package size, all due to the use of a dedicated battery. The major advantage of passive RFID tags is that they can operate without a battery, thus offering much lower production cost (usually a few pennies), longer lifetime (20 years or more), and much smaller package size. For many years, the main drawback of passive tags was known to be their limited operating range, e.g., 3–5 m. Recent tags with operating range up to 100 m have been developed [13], thus making them the tags of choice for most RFID applications.

1.2 Challenges in Wireless Sensing

Although wireless temperature sensors seem very promising, there are many challenges associated with their implementation. To be cost-effective, such sensors must be fully compatible with CMOS technology. Fortunately, various temperature sensing elements are available in standard CMOS technology. However, due to the process spread of various elements, CMOS sensors often require sophisticated and/or time-consuming *calibration* and *trimming* processes (e.g., two-temperature calibration and trimming) to obtain sufficient accuracy. The calibration process is usually performed by comparing the sensor's output with that of a reference sensor at a number of known temperatures. Since both sensors need to reach thermal equilibrium, such thermal calibration can take several tens of seconds. The extra time required to perform calibration and trimming, however, increases the production cost, and thus sensors with no calibration or a minimum number of calibration points are desired. Alternatively, calibration techniques based on electrical measurements can be developed to simultaneously achieve low cost and good accuracy [14].

The required accuracy of a temperature-sensing node depends on the target application, ranging from ±0.1 °C for medical [15, 16] to ±1 °C for food and environmental monitoring applications [17]. The operating temperature range also depends on the target application, e.g., from 35 to 45 °C in medical applications, from −40 to 85 °C in environmental monitoring, and from −40 to 150 °C in automotive applications. The actual number of required calibration points then depends on the type of sensing element, the target accuracy, and the sensor's operating temperature range. Clearly, there is a trade-off between the number of required calibration points (and therefore cost) and the target accuracy for a given application.

Furthermore, in the design of temperature-sensing wireless nodes, the power and energy efficiency of the co-integrated temperature sensor are key parameters. Typical CMOS smart sensors suffer from relatively high power consumption, e.g., 500 μA in [3] and 2.2 mA in [5], and/or long conversion time (T_{conv}), e.g., 300 ms in [3] and 1.5 s in [4], which results in high "energy consumption." Such sensors are ill suited for use in battery-powered WSNs or active RFID tags as they would dramatically decrease the battery's lifetime, and thus are not cost-effective. They are also not suitable for use in passive RFID tags or WSNs operating based on energy harvesting or scavenging. This is due to the restricted amount of energy available in such systems, which either limits the maximum communication range or requires a larger antenna or energy storing element, e.g., a capacitor, or calls for using energy harvesters. Moreover, the power received at a passive RFID tag falls off as the square of the distance. Therefore, there is a trade-off between the sensor's energy consumption on the one hand, and the operating range, size, and cost of the sensor node on the other hand. This implies that energy-efficient sensors, i.e., low-power (e.g., a few μW) sensors with fast conversion times are essential for wireless temperature sensing applications.

Temperature sensors for wireless sensing were introduced prior to the start of this research [17, 18]. The design in [17] presents a temperature sensor, which is embedded into a passive-RFID tag. The tag dissipates $10\,\mu$A to operate and requires a conversion time of 510 ms. It achieves an inaccuracy of $\pm 2.5\,°$C (four samples) from 0 to 100 °C, after a one-point calibration. The read range is limited to 10–25 cm, depending on the size of the antenna used. The sensor in [18] is quite power-efficient, dissipating 220 nW from a 1 V supply. However, it requires a conversion time of 100 ms to obtain a resolution of 0.1 °C. Furthermore, it requires a two-point calibration to achieve an inaccuracy of $-1.6\,°$C/$+3\,°$C (five samples) from 0 to 100 °C. In 2010, a sensor was presented which dissipates 100 nW, and achieves a resolution of 35 mK in a conversion time of 100 ms [16]. It also achieves an inaccuracy of $\pm 0.1\,°$C (three samples), over a range from 35 to 45 °C, but only after a two-point calibration. Recently, another temperature sensor embedded into a passive RFID tag has been presented [19]. The sensor dissipates 350 nA from a 1 V supply. After a one-point calibration, it achieves an inaccuracy of $\pm 1.5\,°$C (3σ) from -30 to 60 °C. In a conversion time of 14.5 ms, it obtains a resolution of 0.3 °C. As can be seen, most of these low power/energy sensors suffer from poor accuracy, even after calibration.

In this thesis, we will focus on the design of low-cost, accurate, and energy-efficient CMOS temperature sensors. To understand the existing design trade-offs, we will first review various CMOS-compatible sensing elements from the perspectives of accuracy and energy efficiency, which will be presented in the following section. A general figure-of-merit (FoM) will then be presented, which will facilitate comparisons between the energy efficiency of different types of sensors. Lastly, a short survey of the state of the art in 2009 will be provided, which enables us to evaluate the state of the art at the start of this research.

1.3   CMOS-Compatible Sensing Elements

In CMOS technology, the temperature dependence of several different circuit elements can be used for temperature sensing. The correct choice of sensing element, however, is not trivial and depends on the requirements of the target application, such as accuracy, resolution, power consumption, conversion time, operating supply voltage range, operating temperature range, and power supply rejection ratio (PSRR). In the following, various CMOS-compatible sensing elements are briefly introduced and then investigated based on some of the aforementioned requirements.

1.3.1   Bipolar Junction Transistors (BJTs)

In CMOS technology, the same diffusions normally used to realize MOSFETs can be used to realize *parasitic* vertical bipolar junction transistors (BJTs). While smart temperature sensors based on lateral PNP transistors have been realized [20, 21],

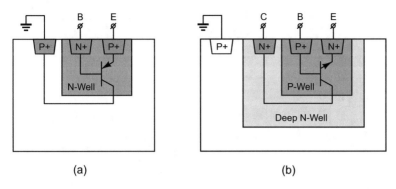

Fig. 1.5 (**a**) Cross section of vertical PNP transistors in standard CMOS; (**b**) cross section of vertical NPN transistors in modern CMOS technology supporting deep N-well

nowadays vertical PNP transistors are preferred due to their lower sensitivity to process spread and packaging stress [22, 23]. Such parasitic vertical PNPs, however, usually offer limited implementation flexibility, collector is formed inside the P-substrate, and thus, is not directly accessible (see Fig. 1.5a). In modern CMOS technologies with twin well or deep N-Well options, vertical NPN transistors are also available as shown in Fig. 1.5b. They exhibit significantly larger current gain than PNPs, e.g., $\beta_F = 24$ (NPNs) versus $\beta_F = 4$ (PNPs) in a TSMC 0.18 μm CMOS technology. NPNs also offer more circuit design flexibility, since their collector terminals are accessible.

The base-emitter voltage V_{BE} of a BJT can be expressed as follows:

$$V_{BE} \approx \frac{kT}{q} \ln \left(\frac{I_C}{I_S} + 1 \right), \tag{1.1}$$

where k, T, and q denote the Boltzmann constant (1.38×10^{-23} J/K), the temperature in Kelvin, and the electron charge (1.6×10^{-19} C), respectively. The parameter I_S denotes the saturation current of the bipolar transistor. It can be shown that V_{BE} exhibits complementary-to-absolute temperature (CTAT) behavior with a slope of ≈ -2 mV/°C [2]. However, if two BJTs are biased at different collector current densities with a ratio p, the difference $\Delta V_{BE} = V_{BE2} - V_{BE1}$ will be a proportional-to-absolute temperature (PTAT) voltage with a temperature coefficient that depends on the constants k/q and the ratio p [2]. The well-defined temperature dependency of V_{BE} and ΔV_{BE} makes BJTs attractive for use in CMOS temperature sensors and bandgap voltage references. In fact, BJT-based temperature sensors have been widely used in the industry for decades [3–7]. The reasons for this are as follows: for a properly designed sensor, the dominant source of inaccuracy is the process spread in V_{BE}, which has been shown to have a PTAT profile [2], and thus can be corrected by means of a cost-effective one-point PTAT trim, e.g., ±0.5 °C (3σ) from −50 to 120 °C in [24] and ±0.1 °C (3σ) from −55 to 125 °C in [25]. Another advantage is that the necessary temperature dependent and reference voltages are both generated

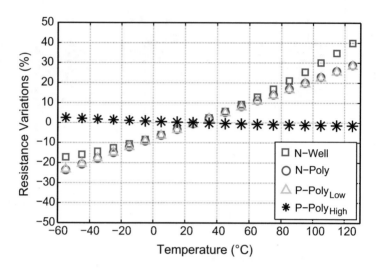

Fig. 1.6 Temperature dependency of some types of resistors available in a TSMC 0.18 μm process. Resistance variations are normalized to the value at 25 °C

from the same circuit, which significantly simplifies the implementation. They require bias currents in the range of μA or even sub-μA to operate, and exhibit low supply dependency, usually a few tenths of degrees Celsius per Volt, e.g., 0.5 °C/V in [24] and 0.1 °C/V in [25].

1.3.2   Resistors

Resistance temperature detectors (RTDs) have been widely used as stand-alone temperature sensing elements. Temperature information is obtained by reading out resistance variations as a function of temperature, implying that a large temperature-coefficient is often desired. As it turns out, most CMOS-compatible resistors exhibit significant temperature coefficients, with 1st-order coefficients ranging between 0.1%/°C and 0.4%/°C, depending on the resistor type. Figure 1.6 shows the simulated temperature dependency of some of the resistors available in the TSMC 0.18 μm CMOS process. The variations are normalized to the resistance at 25 °C. The temperature coefficient of +0.4%/°C exhibited by a typical N-Well or N-Poly resistor means that its resistance will increase by about 72% over the temperature range from −55 to 125 °C, which is reasonably large sensitivity. In such resistor-based sensors, the minimum supply voltage is usually limited by the readout circuit, thus enabling low supply voltages. The value of the bias current is defined by thermal-noise and area constraints.

A drawback of resistors as temperature sensing elements is the fact that the spread of most resistances in CMOS is in the range of 15–20% across the process corners. Their temperature coefficients also suffer from process spread and higher order nonlinear terms, as can be noticed from Fig. 1.6. As a result, resistors usually require a costly multiple-temperature calibration to achieve decent accuracy, where the number of calibration points could range between 3 and 5, depending on the target accuracy. The work presented in [26] and [27], for example, both achieve an inaccuracy of $\pm 0.15\,^\circ$C (3σ) from -55 to $85\,^\circ$C, but only after a costly three-temperature trim. Employing a single temperature trim, the work in [28] achieves an inaccuracy of $\pm 1\,^\circ$C (3σ) from -45 to $125\,^\circ$C, which is among the best reported for similarly trimmed resistor-based sensors.

1.3.3 Electro-Thermal Filters (ETFs)

The thermal diffusivity of silicon D is defined as the rate at which heat diffuses through a silicon substrate. Recent research has shown that D is a well-defined parameter, as the silicon used for IC fabrication is highly pure [29]. Furthermore, D is strongly temperature dependent and can be approximated by a power law: $D \propto 1/T^{1.8}$ [30–32]. This well-defined temperature dependency can thus be exploited to realize temperature sensors. Figure 1.7 shows the structure of an electro-thermal filter (ETF), which uses a heater to generate heat pulses, and a (relative) temperature sensor (thermopile), fabricated at a distance s from the heater, which converts the received temperature variations into a small voltage signal. In the thermal domain, an ETF behaves like a low-pass filter. Driving such a filter at a given excitation frequency results in a temperature-dependent phase-shift [32, 33]:

$$\phi_{\text{ETF}} \propto (s\sqrt{f_{\text{ref}}})T^{n/2}, \tag{1.2}$$

where $n \approx 1.8$. A phase-domain ADC can then be used to digitize ϕ_{ETF} and obtain temperature in digital format [33]. Figure 1.8 shows the phase-shift ϕ_{ETF} versus temperature for a typical ETF. As shown, and is also clear from the above expression, ϕ_{ETF} is slightly nonlinear with temperature, which calls for linearization in the digital domain.

Since an ETF requires heat pulses to operate, it is naturally ill suited to low-power applications, e.g., the ETF-based sensor presented in [33] requires 5 mW to operate. However, decent accuracies can be obtained without trimming, and only based on batch-calibration of sensors, e.g., $\pm 0.5\,^\circ$C (3σ) from -55 to $125\,^\circ$C in $0.7\,\mu$m

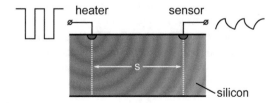

Fig. 1.7 Cross section of an electro-thermal filter (ETF) consisting of a heater and a temperature sensor (thermopile) at a distance s formed in the silicon substrate

Fig. 1.8 Phase shift of an electro-thermal filter (ETF) as a function of temperature [33]

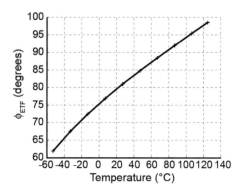

CMOS process [33], and even $\pm 0.2\,^{\circ}\text{C}$ (3σ) in $0.18\,\mu\text{m}$ CMOS [34]. This is due to the fact that the accuracy of ETF-based sensors depends on that of the lithography that realizes the distance s, and is thus expected to scale with every CMOS process node. This makes such sensors quite promising in applications where uncalibrated accuracy is critical, while their relatively large power consumption can be tolerated, e.g., in the thermal management of microprocessors.

1.3.4 MOSFETs

When biased in the sub-threshold region, the drain current I_D and the gate-source voltage V_{GS} of a MOSFET exhibit a temperature-dependent exponential relationship, similar to that between the collector current I_C and V_{BE} of a BJT [35]:

$$I_D^{\text{bulk}} \propto \frac{W}{L} \exp\left[\frac{q}{mkT}(V_{\text{GS}} - V_T^{\text{bulk}})\right], \tag{1.3}$$

where k is the Boltzmann's constant, T is the absolute temperature, and q is the electron charge, and W and L represent the width and length of the device, respectively. The parameter $m = 1 + C_D/C_{\text{OX}}$, is the body effect coefficient, where C_D and C_{OX} are the depletion-layer and gate-oxide capacitances, respectively [35]. Similar exponential relationships between Eqs. (1.1) and (1.3) suggest that MOSFETS can replace BJTs as temperature sensing elements [36]. Compared to BJTs, however, the gate-source voltage V_{GS} of a MOSFET biased in sub-threshold is substantially smaller and can be controlled by sizing W and/or L. This, in turn, offers a potential advantage for low supply voltage operation. However, the oxide capacitance C_{OX} suffers from process spread, while the threshold voltage V_T^{bulk} also varies due to the body-effect and suffers from the process spread as well. In consequence, MOSFET-based sensors suffer from the process spread of two different parameters, which, in turn, results in greater inaccuracies when compared to equally one-point calibrated BJTs. Therefore, MOSFET-based sensors often

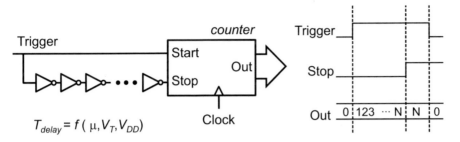

Fig. 1.9 Block diagram of a MOSFET-based temperature sensor based on inverter delay

require two-temperature calibration to meet the accuracy requirements of most of the applications.

The propagation delay of a CMOS inverter chain, or alternatively, the frequency of a ring oscillator, can also be used as a measure of temperature [37]. Figure 1.9 shows the operating principle of such sensors, where a counter is used to measure the propagation delay through a chain of inverters. The average propagation delay T_P of an inverter composed of balanced PMOS and NMOS devices can be expressed as [37]:

$$T_P = \frac{(L/W)C_L}{\mu C_{OX}(V_{DD} - V_T)} \cdot \ln\left[\frac{3V_{DD} - 4V_T}{V_{DD}}\right], \tag{1.4}$$

in which the mobility μ and V_T are temperature-dependent parameters. Assuming $V_{DD} \gg V_T$, then T_P will depend on temperature mainly through μ. This assumption, however, becomes less and less valid in the modern CMOS processes with reduced supply voltages. Besides, T_P suffers from the process spread in V_T and from the variations in V_{DD} as well. In consequence, such sensors usually require two-point calibration and suffer from a poor power supply sensitivity, usually in the range of several degrees Celsius per Volt, e.g., $10\,°C/V$ in [38]. This is about two orders of magnitude worse than typical BJT-based sensors and is prohibitively large for most of the applications. Therefore, in practice, such sensors should be used with voltage regulators, which calls for extra area and power consumption.

1.3.5 Dynamic Threshold MOSFETs (DTMOSTs)

Consider a standard MOSFET biased in sub-threshold region, with the gate and bulk terminals tied together, as shown in Fig. 1.10. This connection fixes the width of the depletion layer under the channel, thereby causing the threshold voltage to vary dynamically, hence the name dynamic-threshold MOST (DTMOST). As a result, the drain current I_D^{DT} of a DTMOS transistor operated in the sub-threshold region can be expressed as follows [39]:

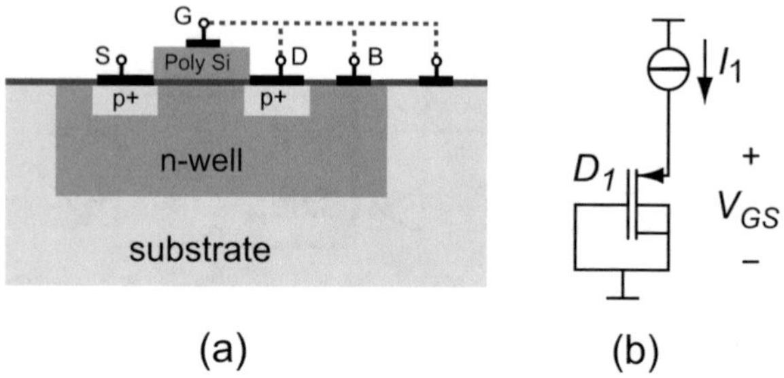

Fig. 1.10 A P-type
DTMOST diode; cross
section view (**a**), symbol
view (**b**)

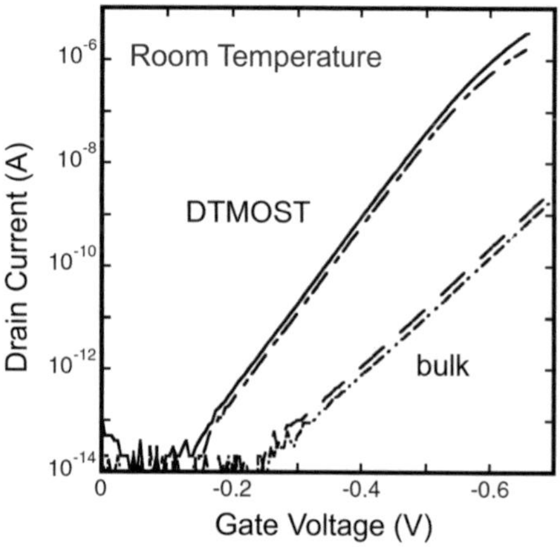

(a)   (b)

Fig. 1.11 Subthreshold
characteristics of a bulk
PMOS device operated in
both "bulk" and "DTMOST"
modes, measured at room
temperature [39]

$$I_D^{\mathrm{DT}} \propto \frac{W}{L} \exp\left[\frac{q}{kT}(V_{\mathrm{GS}} - V_T^{\mathrm{DT}}) \right], \tag{1.5}$$

The key observation is that the gate-body connection ensures that the threshold voltage V_T^{DT} of a DTMOS transistor is well defined. As a result, a diode-connected DTMOST, i.e., a DTMOS diode exhibits a near-ideal exponential relationship between I_D^{DT} and V_{GS}, which is less dependent on C_{OX} and C_D [35, 39]. Figure 1.11 compares the sub-threshold characteristics of a bulk PMOST operated in both *bulk* mode (gate and substrate electrically isolated) and DTMOST mode. As shown, a DTMOST configuration would result in a steeper sub-threshold slope, lowered threshold voltage, and thus higher I_D, when compared to the bulk configuration for the same device.

More importantly, unlike the bulk configuration, the sub-threshold slope in the DTMOST configuration is well defined and is less dependent on device-related parameters, as can be also seen from Eq. (1.5). In other words, the process spread of V_{GS} in the DTMOST configuration is less than that of the bulk configuration [39, 40]. This would suggest that similar to BJTs, DTMOSTs can be effectively calibrated at a single temperature, while offering the low-voltage capability of MOSFETs.

1.4 Energy Efficiency and Resolution FoM

Given the variety of smart temperature sensors in standard CMOS, devising a single figure of merit (FoM) to assess their energy efficiency performance would be very useful. As shown in Fig. 1.1, a smart temperature sensor typically consists of an ADC that digitizes the sensor's front-end output: usually a small signal contaminated by the thermal-noise. Moreover, the resolution of an optimally designed ADC is limited by thermal- rather than quantization-noise. This would suggest that as for general-purpose ADCs [41], a resolution figure-of-merit (FoM) involving the energy per conversion and resolution could be defined as follows [42]:

$$\text{FoM} = E_{\text{conv}} \cdot \text{Resolution}^2, \tag{1.6}$$

where E_{conv} is the amount of energy dissipated per conversion. It should be noted that in the context of smart temperature sensors, other figures of merit involving the sensor's accuracy might also be useful. However, this is complicated by the fact that various sensors employ different numbers of calibration points, making a fair comparison rather difficult.

Figure 1.12 shows the energy per conversion versus the resolution of several smart temperature sensors, published prior to the start of this research (in 2009). It can be seen that the resolution FoM defines a line that usefully bounds the state of the art, as would be expected for thermal-noise limited converters.

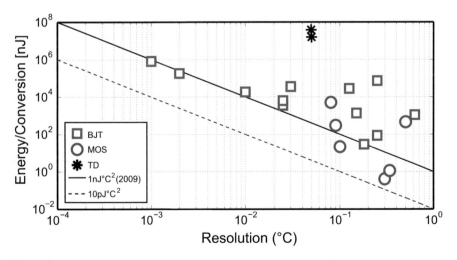

Fig. 1.12 Energy per conversion versus the resolution of several smart temperature sensors, published prior to the start of this research (in 2009) [42]. Note that no resistor-based sensors were published before 2009

1.5 Prior-Art and Choice of Sensing Element

Table 1.1 lists some prior-art temperature sensors (back in 2009), employing different types of sensing elements. As shown, the work in [25] presents a PNP-based temperature sensor which draws 75 μA from a 2.5 V supply. Thanks to the well-defined characteristics of PNPs in the used 0.7 μm CMOS, and combined with various dynamic correction techniques, it achieves an excellent inaccuracy of ±0.1 °C (3σ, 24 samples) over the military temperature range (−55 to 125 °C) and employing a single temperature trim. In a conversion time of 100 ms it achieves a thermal noise limited resolution of 10 mK, resulting in a FoM of 1.9 nJ °C². For the targeted energy efficiency in this work, however, the current consumption of 75 μA is prohibitively large and cannot be afforded in wireless temperature sensing applications. Moreover, it uses a rather old process (0.7 μm CMOS), resulting in a large die area of 4.5 mm². The work in [43] shows that similar accuracy can be achieved even after reducing the bias current of PNPs to 100 nA, i.e., by 10× compared to [25]. This suggests that PNPs are capable of achieving a decent accuracy over a wide range of bias current and using a single temperature trim.

The work in [44] presents a sensor based on MOSFET transistors operating in the linear region. It is quite power-efficient, drawing only 400 nA from a 1 V supply. In a conversion time of 1 ms, it achieves a resolution of 300 mK, thus

Table 1.1 Performance summary of prior-art temperature sensors, back in 2009

Parameter	JSSC'05 [25]	ISSCC'09 [25]	TCAS'09 [44]	JSSC'05 [37]	ISSCC'08 [33]
Sensor type	BJT	BJT	MOSFET	MOSFET	Thermal diffusivity
CMOS technology	0.7 μm	0.7 μm	0.18 μm	0.35 μm	0.7 μm
Chip area	4.5 mm²	4.5 mm²	0.032 mm²	0.175 mm²	2.3 mm²
Supply current	75 μA	25 μA	0.4 μA	3 μA	1 mA
Min supply voltage	2.5 V	2.5 V	1 V	3.3 V	5 V
Supply sensitivity	0.03 °C/V	0.05 °C/V	8 °C/V	0.1 °C/V	–
Inaccuracy (calibration points)	±0.1 °C(3σ) (1)	±0.1 °C (3σ) (1)	−0.8/+1 °C (2)	−0.7/+0.9 °C (2)	±0.5 °C(3σ) (batch-calibrated)
Temperature range	−55 to 125 °C	−55 to 125 °C	0 to 100 °C	0 to 100 °C	−55 to 125 °C
Resolution T_{conv}	0.01 °C (100 ms)	0.025 °C (100 ms)	0.3 °C (1 ms)	0.08 °C (500 ms)	0.05 °C (3030 ms)
Resolution FoM	1.9 nJ °C²	3.9 nJ °C²	36 pJ °C²	32 nJ °C²	38 μJ °C²

obtaining a decent energy efficiency evidenced by a FoM of 36 pJ $°C^2$. It is also quite compact, occupying only 0.032 mm^2 in a 0.18 μm CMOS process. However, over the temperature range from 0 to 100 °C it achieves an inaccuracy of $-0.8/+1$ °C after a costly two-end-point trim (four samples). Besides, it suffers from a poor supply dependency of 8°/V. In [37], a sensor based on the temperature dependence of delay lines is presented. For the class of MOSFET-based designs, it achieves the best inaccuracy of $-0.7/+0.9$ °C over the temperature range from 0 to 100 °C, but only after a two-end-point trim (25 samples). It also requires a long conversion time to operate; 500 ms to achieve a resolution of 80 mK, while drawing 3 μA from a 3.3 V supply. This results in a poor energy efficiency of 32 nJ $°C^2$.

The work in [33] presents an ETF-based temperature sensor. It achieves an untrimmed (but batch-calibrated) inaccuracy of $±0.5$ °C ($3σ$, 16 samples) over the military temperature range (-55 to 125 °C). The fact that a decent accuracy can be achieved without trimming makes ETF-based sensors very attractive, especially for low-cost applications, in which calibration cost needs to be minimized. Nevertheless, as previously mentioned, such sensors are rather power-hungry, e.g., drawing 1 mA from a 5 V supply in [33].

Studying the different sensors in Table 1.1 reveals the fact that BJTs are the devices of choice for wireless temperature sensing applications, in which energy efficiency, low-cost, and precision are key requirements. Their bias current can be reduced to the tens of nA level with minimal impact on the accuracy. They also exhibit a well-defined process spread which can be effectively trimmed at a single temperature. None of the MOSFET- or ETF-based sensors can offer such a combination. Back in 2009, no resistor- or DTMOST-based sensors existed, and thus these types are missing in Table 1.1. It is, however, well understood that resistor-based sensors, although potentially suitable for use as energy-efficient sensors, require multiple-temperature calibration which rules out their use for low-cost applications. DTMOSTs, on the other hand, seem to be capable of offering high accuracy using a single temperature trim. In order to operate in weak inversion, their bias current is naturally in the sub-μA range, thus offering low-power operation as well. A distinct advantage of DTMOSTs compared to BJTs would be their capability for sub-1 V supply operation. However, no experimental DTMOST-based sensor prototype had been realized prior to this research and thus no apples-to-apples comparison could be made.

1.6 Thesis Organization

In the next chapter, an architecture-level solution to realize energy-efficient BJT-based sensors is presented. This is done by analyzing existing sensor architectures and finding out the root causes of their lack of energy efficiency. To bridge the energy efficiency gap between the existing solutions and requirements of the target application, a novel *zoom*-ADC architecture will be presented which results in low-power and fast conversions, thereby significantly improving the energy efficiency. This, in turn, results in a compact, and therefore, low-cost solution. In Chap. 3, two sensor prototypes based on 1st- and 2nd-order zoom-ADCs will be presented. Moreover, another prototype is presented for use in automotive applications, where sensing high temperatures ($>150\,^\circ$C) is desired. In this research, DTMOST-based sensors have also been realized, which is the subject of Chap. 4. First, a DTMOST-based prototype based on a 1st-order zoom-ADC will be presented and an apples-to-apples comparison to BJT-based sensors will be made. It is shown that DTMOSTs are in fact capable of achieving high accuracy after a single temperature trim, and so can bridge the accuracy gap between the MOSFET- and BJT-based sensors. Based on the low-voltage operation capability of DTMOSTs, a second prototype is then realized which provides both sub-1 V, and sub-μA operations. Chapter 5 is devoted to the conclusions and a comparison between the different sensor prototypes realized in this research and state of the art.

References

1. National Semiconductor, National Semiconductor Temperature Sensor Handbook, http://www.shrubbery.net/~heas/willem/PDF/NSC/temphb.pdf
2. M.A.P. Pertijs, J.H. Huijsing, *Precision Temperature Sensors in CMOS Technology* (Springer, Dordrecht, 2006)
3. Texas Instruments Inc., LM75 data sheet, July 2009, www.ti.com
4. Maxim Int. Prod., DS1775 data sheet, May 2013, www.maxim-ic.com
5. Analog Devices Inc., ADT7301 data sheet, Aug. 2005, www.analog.com
6. NXP Semiconductors, PCT2075 data sheet, Oct. 2014, www.nxp.com
7. NXP Semiconductors, PCT2202UK data sheet, Aug. 2015, www.nxp.com
8. K. Sohraby, D. Minoli, T. Znati, *Wireless Sensor Networks: Technology, Protocols, and Applications* (Wiley, Hoboken, 2007), pp. 203–209. ISBN 978-0-471-74300-2
9. Cymbet Corp., http://www.cymbet.com
10. http://www.statista.com/statistics
11. https://www.smartrac-group.com
12. http://www.icinsights.com/news/bulletins
13. Mojix Inc., http://www.mojix.com
14. M.A.P. Pertijs, A.L. Aita, K.A.A. Makinwa, J.H. Huijsing, Voltage calibration of smart temperature sensors, in *Proceedings of IEEE Sensors*, Oct. 2008, pp. 756–759
15. G.C.M. Meijer, A.J.M. Boomkamp, R.J. Duguesnoy, An accurate biomedical temperature transducer with on-chip microcomputer interfacing. IEEE J. Solid State Circuits **23**(6), 1405–1410 (1998)
16. A. Vaz et al., Full passive UHF tag with a temperature sensor suitable for human body temperature monitoring. IEEE Trans. Circuits Syst. II **57**(2), 95–99 (2010)
17. K. Opasjumruskit et al., Self-powered wireless temperature sensors exploit RFID technology. IEEE Pervasive Comput. **5**, 54–61 (2006)
18. Y.S. Lin, D. Sylvester, D. Blaauw, An ultra low power 1V, 220 nW temperature sensor for passive wireless applications, in *Proceedings of CICC*, Sept. 2008, pp. 507–510
19. B. Wang, M.K. Law, A. Bermark, H.C. Luong, A passive RFID tag embedded temperature sensor with improved process spreads immunity for a -30°C to 60°C sensing range. IEEE Trans. Circuits Syst. **61**(2), 337–346 (2014)

20. P. Krummenacher, H. Oguey, Smart temperature sensors in CMOS technology. Sens. Actuators A **22**(1–3), 636–638 (1990)
21. R.A. Bianchi et al., CMOS-compatible temperature sensor with digital output for wide temperature range applications. Microelectron. J. **31**, 803–810 (2000)
22. J. F. Creemer, F. Fruett, G.C. Meijer, P.J. French, The piezojunction effect in silicon sensors and circuits and its relation to piezoresistance. IEEE Sens. J. **1**(2), 98–108 (2001)
23. F. Fruett, G.C. Meijer, *The Piezojunction Effect in Silicon Integrated Circuits and Sensors* (Kluwer Academic, Boston, 2002)
24. M.A.P. Pertijs et al., A CMOS temperature sensor with a 3σ inaccuracy of $\pm 0.5°$C from $-50°$C to $120°$C. IEEE J. Solid State Circuits **40**(2), 454–461 (2005)
25. M.A.P. Pertijs, K.A.A. Makinwa, J.H. Huijsing, A CMOS temperature sensor with a 3σ inaccuracy of $\pm 0.1°$C from $-55°$C to $125°$C. IEEE J. Solid State Circuits **40**(12), 2805–2815 (2005)
26. M. Shahmohammadi, K. Souri, K.A.A. Makinwa, A resistor-based temperature sensor for MEMS frequency references, in *Proceedings of ESSCIRC*, Sept. 2013, pp. 225–228
27. P. Park, K.A.A. Makinwa, D. Ruffieux, A resistor-based temperature sensor for a real time clock with ± 2ppm frequency stability, in *Proceedings of ESSCIRC*, Sept. 2014, pp. 391–394
28. C.-H. Weng et al., A CMOS thermistor-embedded continuous-time delta-sigma temperature sensor with a resolution FoM of 0.65pJ$°$C^2. IEEE J. Solid State Circuits **50**(11), 2491–2500 (2015)
29. T. Veijola, Simple model for thermal spreading impedance, in *Proceedings of BEC*, Oct. 1996, pp. 73–76
30. Y.S. Touloukian et al., *Thermophysical Properties of Matter*, vol. 10 (Plenum, New York, 1998)
31. S.M. Kashmiri, K.A.A. Makinwa, Measuring the thermal diffusivity of CMOS chips, in *Proceedings of IEEE Sensors*, Oct. 2009, pp. 45–48
32. K.A.A. Makinwa, M.F. Snoeij, A CMOS temperature-to-frequency converter with an inaccuracy of $\pm 0.5°$C (3σ) from $-40°$C to $105°$C. IEEE J. Solid State Circuits **41**(12), 2992–2997 (2006)
33. C.P.L. van Vroonhoven, K.A.A. Makinwa, A CMOS temperature-to-digital converter with an inaccuracy of $\pm 0.5°$C (3σ) from $-55°$C to $125°$C, in *Digest of Technical Papers ISSCC*, Feb 2008, pp. 576–577
34. C.P.L. van Vroonhoven, D. d'Aquino, K.A.A. Makinwa, A thermal-diffusivity-based temperature sensor with an untrimmed inaccuracy of $\pm 0.2°$C (3σ) from $-55°$C to $125°$C, in *Digest of Technical Papers ISSCC*, Feb 2010, pp. 314–315
35. M. Terauchi, Selectable logarithmic/linear response active pixel sensor cell with reduced fixed-pattern-noise based on dynamic threshold MOS operation. Jpn. J. Appl. Phys. **44**(4B), 2347–2350 (2005)
36. K. Ueno et al., Ultralow-power smart temperature sensor with subthreshold CMOS circuits, in *Proceedings of the International Symposium on Intelligent Signal Processing and Communications (ISPACS)*, Dec. 2006, pp. 546–549
37. P. Chen et al., A time-to-digital-converter-based CMOS smart temperature sensor. IEEE J. Solid State Circuits **40**(8), 1642–1648 (2005)
38. P. Chen et al., A time-domain SAR smart temperature sensor with a 3σ inaccuracy of $-0.4°$C $\sim +0.6°$C over a $0°$C to $90°$C range. IEEE J. Solid State Circuits **45**(3), 600–609 (2010)
39. M. Terauchi, Temperature dependence of the subthreshold characteristics of dynamic threshold MOSFETs and its application to an absolute-temperature sensing scheme for low-voltage operation. Jpn. J. Appl. Phys. **46**(7A), 4102–4104 (2007)
40. A.J. Annema, Low-power bandgap references featuring DTMOS. IEEE J. Solid State Circuits **34**(72), 949–955 (1999)
41. B. Murmann, ADC Performance Survey 1997–2015, [Online]. Available: http://web.stanford.edu/~murmann/adcsurvey.html
42. K.A.A. Makinwa, Smart Temperature Sensor Survey, [Online]. Available: http://ei.ewi.tudelft.nl/docs/TSensor_survey.xls

43. A.L. Aita, M.A.P. Pertijs, K.A.A. Makinwa, J.H. Huijsing, A CMOS smart temperature sensor with a batch-calibrated inaccuracy of $\pm 0.25°C$ (3σ) from $-70°C$ to $130°C$, in *Digest of Technical Papers ISSCC*, Feb 2009, pp. 342–343
44. M.K. Law, A. Bermak, A 405-nW CMOS temperature sensor based on linear MOS operation. IEEE Trans. Circuits Syst.-II **56**, 891–895 (2009)

Chapter 2
Readout Methods for BJT-Based Temperature Sensors

2.1 Introduction

As discussed in the previous chapter, BJT-based temperature sensors are promising candidates for use in wireless temperature sensing applications. In this chapter, we first describe the operating principle of BJT-based sensors, followed by an overview of various readout methods. The energy efficiency of these methods is then discussed and compared to the ultimate achievable efficiency of BJT-based sensors.

2.2 Operating Principle of BJT-Based Sensors

In general, a digital representation of temperature can be obtained by generating a temperature-dependent voltage and a temperature-independent or reference voltage V_{REF} and then digitizing the ratio with an analog-to-digital converter (ADC). In BJT-based temperature sensors, the bandgap voltage of silicon $V_{bg} \approx 1.2$ V serves as the reference voltage, and hence they are often known as *bandgaptemperaturesensors*.

A smart temperature sensor therefore requires both a well-behaved temperature-dependent signal, preferably one that is linearly proportional-to-absolute temperature (PTAT), as well as a reference voltage V_{REF} for temperature readout. The main advantage of BJTs compared to other temperature sensing elements is that they can be used to generate both V_{PTAT} and V_{REF} simultaneously and with minimal circuit overhead. Moreover, the output of the resulting temperature sensor is a linear function of temperature, and thus no extra post processing is required. Lastly, but perhaps most importantly for industrial applications, single temperature trimming is enough to achieve high precision, e.g., the $\pm 0.1\,°C$ inaccuracy achieved in [1, 2].

© Springer International Publishing AG 2018   19
K. Souri, K.A.A. Makinwa, *Energy-Efficient Smart Temperature Sensors in CMOS Technology*, Analog Circuits and Signal Processing,
DOI 10.1007/978-3-319-62307-8_2

Fig. 2.1 Diode-connected
configuration of a PNP
transistor [5]

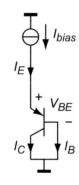

In the 160 nm process (CMOS-14) used in this work, the most suitable BJT is
a substrate-PNP. Compared to the lateral NPN which is also available, it is less
sensitive to process spread and packaging stress [3, 4]. However, its main drawback
is that its collector is grounded, which in turn limits the number of practical circuit
topologies. In fact, the best way of using the substrate PNP is in a diode-connected
configuration, as shown in Fig. 2.1, in which the base-collector junction of the device
is short circuited ($V_{BC} = 0$) [5]. This reduces leakage currents via the base-collector
junction to negligible levels. It also reduces errors due to the transistor's finite
Early voltage. Such leakage currents would otherwise disturb the collector current
accuracy, which, in turn, would impact the sensor's ultimate accuracy.

2.2.1 Temperature Characteristics of BJTs

Figure 2.2a illustrates two bipolar transistors Q_1 and Q_2 (with emitter area of $r \cdot A_E$
and A_E, respectively), configured in a "diode-connected" fashion, biased by currents
I_1 and $p \cdot I_1$. For a diode-connected bipolar transistor, the collector current I_C can be
expressed as [5]:

$$I_C = I_S \left(\exp \left(\frac{q V_{BE}}{kT} \right) - 1 \right), \tag{2.1}$$

where k, T, and q represent the Boltzmann constant (1.38×10^{-23} J/K), temperature
in Kelvin, and electron charge (1.6×10^{-19} C), respectively. V_{BE} is the base-emitter
voltage difference, and the parameter I_S denotes the saturation current of the diode-
connected bipolar transistor and is given by [5]:

$$I_S = CT^\eta \exp \left(\frac{-q V_{g0}}{kT} \right). \tag{2.2}$$

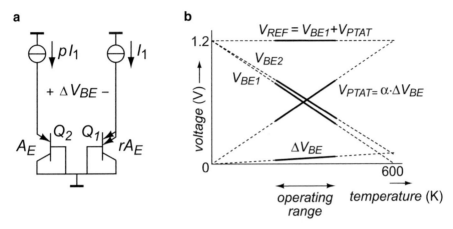

Fig. 2.2 (**a**) Two diode-connected PNPs biased at a collector current density ratio of *pr*; (**b**) these transistors are used to generate V_{PTAT} and V_{REF}, the ratio of which is a measure of temperature [5]

In this equation, C is a constant, V_{g0} is the extrapolated bandgap voltage of silicon at $0\,K$, and $\eta \approx 4$ is a constant for PNP transistors in CMOS technology [6]. We observe that the saturation current I_S is proportional to T^4, implying a strong sensitivity to temperature.

By solving Eq. (2.1), the following expression can be obtained for V_{BE}:

$$V_{\mathrm{BE}} = \frac{kT}{q} \ln \left(\frac{I_C}{I_S} + 1 \right). \tag{2.3}$$

At room temperature, kT/q is about $26\,mV$. Although V_{BE} would seem to have a positive temperature coefficient, it actually has a negative temperature coefficient (about $-2\,mV/^\circ C$) due to the strong temperature dependency of the saturation current I_S. In other words, V_{BE} is, to first order, complementary-to-absolute temperature (CTAT). Figure 2.2b shows how V_{BE} rolls off from $\approx 1.2\,V$ to zero over a temperature range of almost 600°

For two bipolar transistors, e.g., Q_1 and Q_2, biased at different collector current densities, the difference between their base-emitter voltages is a PTAT voltage:

$$\Delta V_{\mathrm{BE}} = V_{\mathrm{BE2}} - V_{\mathrm{BE1}} = \frac{kT}{q} \ln(pr), \tag{2.4}$$

in which r is the emitter area ratio between two bipolar transistors, and p is the collector current ratio (see Fig. 2.2a). As shown in Fig. 2.2b, ΔV_{BE}, unlike V_{BE}, has a positive temperature coefficient which depends on the constants k/q and the product pr. In the rest of this chapter, we will assume the use of two identical bipolar transistors, i.e., $r = 1$, for simplicity.

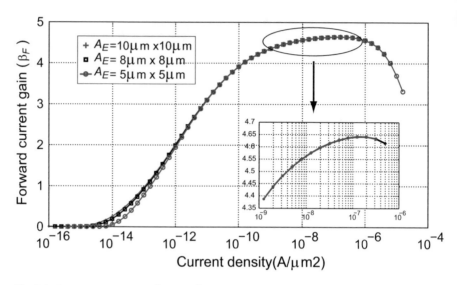

Fig. 2.3 Forward current gain β_F (at 25 °C) of substrate PNPs as a function of collector current density and for different emitter areas in the CMOS-14 process used

For substrate PNPs, however, it is their emitter current ratio, rather than their collector current ratio, that is under control. These ratios are equal if the PNP's forward current gain β_F can be assumed to be constant, i.e., independent of emitter current. This assumption, however, is only valid for older CMOS processes, e.g., 0.7 and 0.5 μm, where β_F is constant for at least two decades of emitter current [1, 7]. Figure 2.3 illustrates the forward current gain β_F as a function of emitter current for substrate PNPs in the target CMOS-14 process. We clearly observe that no flat region exists, thus implying that the collector current ratio is only approximately equal to p [8]. The impact of this is discussed in detail in [8], and is thus neglected in the rest of this chapter.

From a linear combination of V_{BE} (CTAT) and ΔV_{BE} (PTAT), a reference voltage V_{REF} with a nominally zero temperature-coefficient can be generated. As shown in Fig. 2.2b, this can be done by scaling ΔV_{BE} with a constant α such that the positive temperature-coefficient of $\alpha \cdot \Delta V_{BE}$ compensates for the negative temperature-coefficient of V_{BE}:

$$V_{REF} = V_{BE} + \alpha \cdot \Delta V_{BE}. \tag{2.5}$$

Since $\Delta V_{BE} = 0$ at $T = 0$ K, V_{REF} is equal to the extrapolated value of V_{BE} at zero Kelvin, which is related to the bandgap voltage of silicon ($V_{bg} \approx 1.2$ V). Notably, both of the signals required for a ratiometric temperature measurement (V_{PTAT} and V_{REF}) can be obtained by biasing BJTs at different collector current densities. The scaling factor α can be found by solving Eq. (2.5) such that the temperature coefficient of V_{REF} is zero:

$$S^T_{V_{BE}} = \left| \frac{\partial V_{BE}}{\partial T} \right| \approx 2\,\text{mV}/°\text{C} = \alpha \cdot \frac{k}{q} \cdot \ln(p), \qquad (2.6)$$

indicating that:

$$\alpha \approx \frac{S^T_{V_{BE}}}{\frac{k}{q}\ln(p)}. \qquad (2.7)$$

The larger the collector current density ratio p is, the smaller the scaling factor α becomes. Practical values of p typically range between 2 and 20, which means that α typically ranges between 7 and 33. It should be noted that since $S^T_{V_{BE}}$ is a weak function of I_C, as can be seen from Eq. (2.3), α can also be tuned by adjusting I_C.

2.3   Generic BJT Readout

2.3.1   Topology

A block diagram of a generic bandgap temperature sensor is shown in Fig. 2.4. In the analog front-end, two diode-connected bipolar devices Q_1 and Q_2 are biased at a bias current density ratio of p to generate the PTAT signal ΔV_{BE}. The reference voltage V_{REF} then can be obtained by scaling up ΔV_{BE} and adding to V_{BE1}. Finally $\alpha \cdot \Delta V_{BE}$ and V_{REF} voltages are fed to an analog-to-digital converter (ADC) for temperature reading. The ADC then outputs the ratio μ [5]:

$$\mu = \frac{\alpha \cdot \Delta V_{BE}}{V_{BE} + \alpha \cdot \Delta V_{BE}} = \frac{V_{PTAT}}{V_{REF}}, \qquad (2.8)$$

which is PTAT and varies between 0 and 1 over a temperature range of about 600 K. The final step is to linearly scale μ to obtain a digital output D_{out} in degrees Celsius [5]:

$$D_{out} = A \cdot \mu + B, \qquad (2.9)$$

where $A \approx 600\,\text{K}$ and $B \approx -273\,\text{K}$.

Combining V_{BE} and ΔV_{BE} to generate V_{REF}, as shown in Fig. 2.4, requires circuitry and thus power dissipation. As presented in [1], this can be avoided by an alternative scheme in which V_{BE} and ΔV_{BE} are directly input to a $\Delta\Sigma$-ADC. The charge balancing in the ADC then ensures that its output is the desired μ, i.e., V_{REF} is implicitly generated in a dynamic manner. The result is a very simple sensor front-end.

As the PTAT ratio μ lies between 1/3 and 2/3 at the two extremes of the full military temperature range (-55 to $125\,°\text{C}$), only about 30% of the ADC's dynamic

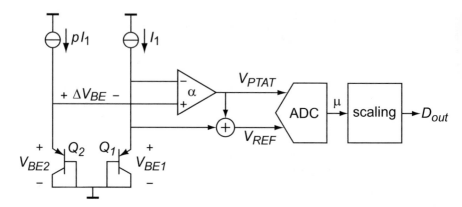

Fig. 2.4 Operating principle of generic bandgap temperature sensors: two diode-connected PNPs generate ΔV_{BE} and V_{BE}. ΔV_{BE} is then amplified and combined with V_{BE1} to provide V_{PTAT} and V_{REF} at the input of an ADC for a ratio-metric measurement [5]

range is used in this configuration. This is also evident from Fig. 2.2b. In other words, the required resolution of the ADC has to be about three times larger than the desired temperature resolution, which means that about 70% of the ADC's dynamic range is wasted. In many other applications, the target temperature range is even smaller, e.g., from -40 to $85\,^{\circ}\mathrm{C}$ and from 25 to $45\,^{\circ}\mathrm{C}$ for industrial and clinical applications, respectively, which further exacerbates the situation.

Another combination of V_{BE} and ΔV_{BE} can be employed to efficiently utilize the ADC's dynamic range and avoid the wasted resolution. As presented in [9, 10], a linear combination of V_{BE} and ΔV_{BE} can be used as the input to the ADC, leading to the following ratio:

$$\mu' = \frac{2\alpha \cdot \Delta V_{\mathrm{BE}} - V_{\mathrm{BE}}}{V_{\mathrm{BE}} + \alpha \cdot \Delta V_{\mathrm{BE}}}. \tag{2.10}$$

Now, as is evident from Fig. 2.5, almost 90% of the ADC's dynamic range is used by the ratio μ' over the military temperature range. In a recent work [11], a different combination of V_{BE} and ΔV_{BE} is used to efficiently utilize the ADC's dynamic range over the clinical temperature range from 25 to $45\,^{\circ}\mathrm{C}$. Employing the generic readout over such a limited range will result in about 96% waste of the ADC's dynamic range. As presented, V_{BE} and ΔV_{BE} are dynamically combined to obtain a linear ratio which varies by over 60% from 25 to $45\,^{\circ}\mathrm{C}$.

From Eq. (2.8), it can be seen that errors in the scaling factor α will impact the accuracy of μ, and thus the digital temperature reading D_{out}. As this scaling has to be done in the analog domain, precision techniques like dynamic element matching (DEM) are needed to realize precision temperature sensors [5, 12]. In addition, α should be an integer or rational number to enable accurate on-chip implementation [13]. This requirement imposes a restriction on the values of I_C and the collector current density ratio p, through Eq. (2.7).

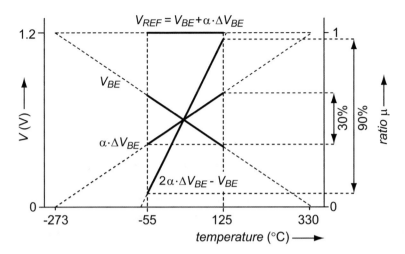

Fig. 2.5 More efficient use of the ADC's dynamic range by using a different combination of V_{BE} and ΔV_{BE} [5]

2.3.2   ADC Resolution

Essentially, the ADC resolution will depend on the target application. Given the diversity of temperature sensing applications, a wide range of resolutions are required, e.g., a couple of hundred milli-Kelvin in environmental monitoring, to sub milli-Kelvin for the temperature compensation of MEMS frequency references. Moreover, the need for accurate calibration (to meet a given accuracy specification) imposes further limits. As a rule of thumb, the sensor's resolution should be an order of magnitude less than the target accuracy to ensure negligible calibration error, e.g., $\pm 0.01\,°C$ resolution if an accuracy of $\pm 0.1\,°C$ is targeted.

In an ADC, the effective number of bits (ENOB), which expresses the total quantization noise of the ADC as a fraction of its full scale, can be expressed as:

$$\text{ENOB} = \log_2 \left(\frac{D_{FS}}{\max |D_{\text{out}} - D_{\text{out,ideal}}|} \right) - 1, \qquad (2.11)$$

where D_{FS} and $|D_{\text{out}} - D_{\text{out,ideal}}|$ represent the full-scale value and quantization error of the ADC, respectively [5].

For the generic bandgap temperature sensor, D_{FS} can be determined from Fig. 2.2b. As depicted, $V_{REF} \approx 1.2\,V$, which is the full scale value in the voltage domain, and corresponds to a temperature range of about $600\,°C$, i.e., $D_{FS} \approx 600\,°C$. Assuming a target temperature resolution of $\pm 0.01\,°C$ in the calibration phase, an ENOB of 14.9 bits is required, which is not trivial to achieve together with extremely low-energy dissipation. By combining V_{BE} and ΔV_{BE} at the ADC's input as in Fig. 2.5, D_{FS} reduces to $\approx 200\,°C$. This would mean that for a $\pm 0.01\,°C$ resolution, an ENOB of 13.3 bits will be required in the ADC, which is about 1.6 bits less than in the generic approach.

2.4 Energy Efficiency of BJT-Based Sensors

A smart temperature sensor consists of two major blocks; the sensor front-end and the ADC. Since achieving high energy efficiency is the main focus of this research, a careful energy efficiency analysis of existing temperature sensor architectures is essential and will be discussed in the following sections.

2.4.1 Efficiency Limits of a BJT-Based Front-End

Regardless of the chosen ADC topology, the ultimate energy efficiency of a smart temperature sensor is limited to that of its front-end. To calculate this efficiency limit, we assume the circuit diagram of Fig. 2.6. Here, two identical diode-connected PNPs are biased with the bias currents I and $p \cdot I$, mirrored from a bias circuit (not shown). The sensor core generates V_{BE} and ΔV_{BE}, whose ratio is then digitized by means of an ideal ADC, i.e., an ADC with infinite resolution and zero power dissipation.

Recall from Chap. 1 that the resolution figure-of-merit (FoM) is expressed as follows:

$$\text{FoM} = E_{\text{conv}} \cdot \text{Resolution}^2, \qquad (2.12)$$

where E_{conv} represents the dissipated energy per conversion. Neglecting the power dissipated in the bias circuit, E_{conv} can be calculated as follows:

$$E_{\text{conv}} = V_{\text{DD}} \cdot (1 + p) \cdot I \cdot T_{\text{conv}}, \qquad (2.13)$$

where T_{conv} is the sensor's conversion time.

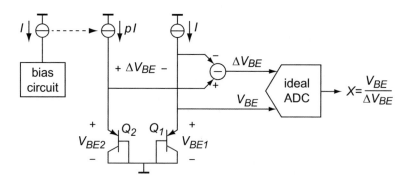

Fig. 2.6 The circuit diagram used to determine the front-end's energy efficiency using bias current scaling in the front-end

The sensor's resolution, on the other hand, is limited by the noise present in the sensor's output voltages, i.e., V_{BE} and ΔV_{BE}. The noise present in the base-emitter voltage of a diode-connected PNP is the combination of thermal noise and shot noise associated with its base resistance and collector current, respectively:

$$v_{n,V_{BE}}^2 = \frac{i_{n,c}^2}{g_m^2} + 4kTR_BB_n = 2qIB_n \left(\frac{kT}{qI}\right)^2 + 4kTR_BB_n$$

$$= \frac{2kT}{g_m}B_n + 4kTR_BB_n, \tag{2.14}$$

where g_m is the PNP's transconductance and B_n is the noise bandwidth. In this analysis the contribution of $1/f$ noise is neglected, as it is relatively small in bipolar transistors. Also, the noise contribution of the base resistance R_B is usually negligible. This is due to the fact that, in practice, PNPs are often biased with current levels in the range of μA or even smaller. The associated $1/g_m$ is, therefore, significantly larger than R_B, and thus dominates the noise in the base-emitter voltage V_{BE}. Assuming that the noise in the bias current I is equally dominated by the shot noise of the PNP transistors of the bias circuit, the overall noise in the base emitter voltage V_{BE} can be formulated as follows:

$$v_{n,V_{BE}}^2 = \frac{2kT}{g_m}B_n + \frac{2qI}{g_m^2}B_n = \frac{4kT}{g_m}B_n. \tag{2.15}$$

The noise present in ΔV_{BE} can then be obtained by adding the noise of two PNPs, as follows:

$$v_{n,\Delta V_{BE}}^2 = v_{n,V_{BE1}}^2 + v_{n,V_{BE2}}^2 = \frac{4kT}{g_m}B_n \cdot \left(1 + \frac{1}{p}\right), \tag{2.16}$$

which is only slightly larger than $v_{n,V_{BE}}^2$.

Assuming that the succeeding (ideal) ADC integrates $v_{n,V_{BE}}^2$ and $v_{n,\Delta V_{BE}}^2$ during a conversion time T_{conv}, which is equivalent to filtering the noise with a sinc filter with a noise bandwidth $B_n = 1/(2 \cdot T_{conv})$, the integrated noise voltage can then be expressed as follows:

$$v_{n,V_{BE}} = \sqrt{\frac{2kT}{g_m} \cdot \frac{1}{T_{conv}}}$$

$$v_{n,\Delta V_{BE}} = \sqrt{\frac{2kT}{g_m} \cdot \frac{1}{T_{conv}} \cdot \left(1 + \frac{1}{p}\right)}. \tag{2.17}$$

The associated noise at the sensor output can now be readily found by calculating the sensitivity of output D_{out} to V_{BE} and ΔV_{BE}. This is achieved by differentiating Eq. (2.9) with respect to V_{BE} and ΔV_{BE} [5]:

$$S_{V_{BE}}^{D_{out}}(T) = \frac{\partial D_{out}}{\partial V_{BE}} = A \cdot \frac{\partial \mu}{\partial V_{BE}} = -\frac{T}{V_{REF}}, \tag{2.18}$$

$$S_{\Delta V_{BE}}^{D_{out}}(T) = \frac{\partial D_{out}}{\partial \Delta V_{BE}} = A \cdot \frac{\partial \mu}{\partial (\Delta V_{BE})} = \frac{A-T}{V_{REF}} \cdot \alpha. \tag{2.19}$$

Clearly, the sensitivity to noise present in ΔV_{BE} is considerably larger than that of V_{BE}. Moreover, given the fact that ΔV_{BE} is always noisier than V_{BE} [see Eq. (2.17)], only the noise contribution of the former will be considered in the following calculations. The noise at the output D_{out} can be computed as:

$$\sigma_T^2 = S_{\Delta V_{BE}}^{D_{out}}(T)^2 \cdot v_{n,\Delta V_{BE}}^2 = \left(\frac{A-T}{V_{REF}}\right)^2 \cdot \alpha^2 \cdot \frac{2kT}{g_m} \cdot \frac{1}{T_{conv}} \cdot \left(1 + \frac{1}{p}\right). \tag{2.20}$$

The front-end FoM can now be calculated by substituting σ_T^2 into Eq. (2.21):

$$FoM = V_{DD} \cdot (1+p) \cdot I \cdot T_{conv} \left(\frac{A-T}{V_{REF}}\right)^2 \cdot \alpha^2 \cdot \frac{2kT}{g_m} \cdot \frac{1}{T_{conv}} \cdot \left(1 + \frac{1}{p}\right), \tag{2.21}$$

which can be simplified to:

$$FoM = 2 \cdot \frac{(p+1)^2}{p} \cdot V_{DD} \cdot q \cdot V_T^2 \cdot \alpha^2 \cdot \left(\frac{A-T}{V_{REF}}\right)^2. \tag{2.22}$$

Surprisingly, the FoM is independent of both the bias current I or the conversion time T_{conv}. This is due to the fact that $E_{conv} \propto (T_{conv} \cdot I)$, while $\sigma_T^2 \propto 1/(T_{conv} \cdot I)$, and so these terms cancel out in the FoM expression. On the other hand, the parameter α depends on p through Eq. (2.7). Therefore, the energy efficiency of the sensor core exclusively depends on p and V_{DD} as design parameters. Figure 2.7 shows the FoM versus temperature and for different p values, assuming $V_{DD} = 1.8\,V$; the nominal supply value in the target $0.16\,\mu m$ CMOS process. As shown, FoM exhibits a parabolic profile with a peak around room temperature. By differentiating Eq. (2.22) with respect to T, it can be shown that FoM reaches to its maximum at $T = A/2 \approx 300\,K$.

Alternatively, the front-end circuit of Fig. 2.8 can be used to generate V_{BE} and ΔV_{BE}. In this circuit two PNPs with an emitter area ratio of $1:p$ are used, each of which is biased with equal bias current of I. The advantage of this biasing scheme is that both BJTs now have the same g_m and hence generate the same noise. The collector current density ratio, essential to develop ΔV_{BE}, is thus achieved by scaling the emitter area, as opposed to scaling the bias current in Fig. 2.6. In this circuit the energy per conversion E_{conv} can be calculated as follows:

$$E_{conv} = 2 \cdot V_{DD} \cdot I \cdot T_{conv}. \tag{2.23}$$

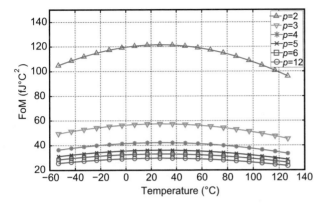

Fig. 2.7 Figure-of-merit (FoM) as a function of temperature and for different p values in the bias current scaling topology

Fig. 2.8 The alternative front-end circuit diagram using scaling of the emitter area

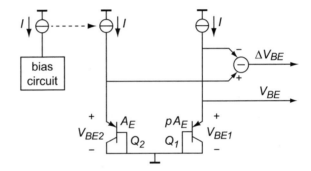

To calculate the FoM of this circuit we first need to find $v^2_{n,\Delta V_{\text{BE}}}$. Neglecting the noise associated with the base resistance R_B, $v^2_{n,\Delta V_{\text{BE}}}$ can be found as:

$$v^2_{n,\Delta V_{\text{BE}}} = v^2_{n,V_{\text{BE1}}} + v^2_{n,V_{\text{BE2}}} \approx \frac{8kT}{g_m} B_n. \tag{2.24}$$

By multiplying by $S^{D_{\text{out}}}_{\Delta V_{\text{BE}}}(T)^2$, the noise at the sensor output σ_T^2 and then the FoM can be calculated as:

$$\text{FoM} = 8 \cdot V_{\text{DD}} \cdot q \cdot V_T^2 \cdot \alpha^2 \cdot \left(\frac{A-T}{V_{\text{REF}}}\right)^2. \tag{2.25}$$

Again, the energy efficiency of front-end depends on V_{DD} and p (via α) as design parameters. Figure 2.9 shows the FoM as a function of p and assuming $V_{\text{DD}} = 1.8$ V. When compared to Fig. 2.7, it can be seen that scaling the emitter area results in a more efficient sensor front-end. This becomes more evident by comparing Eq. (2.22) to (2.25). Clearly, scaling the bias current results in a $(2 + p + 1/p)/4$ times larger

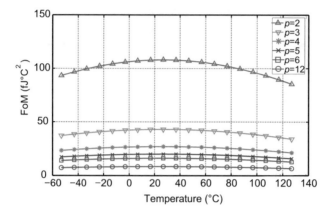

p	Max FoM ($fJ°C^2$)
2	108
3	43
4	27
5	20
6	16
12	8

Fig. 2.9 Figure-of-merit (FoM) as a function of temperature and for different p values in the emitter area scaling topology

Fig. 2.10 Efficiency comparison of bias current versus emitter area scaling as a function of p values

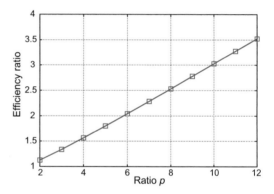

FoM, and thus inferior efficiency. Figure 2.10 illustrates the efficiency comparison as a function of p: the larger the scaling factor p, the less efficient bias current scaling becomes compared to emitter area scaling, e.g., for $p = 6$, bias current scaling is $\approx$ 2× less energy efficient.

Equations (2.22) and (2.25) set the lower bound on the energy efficiency of BJT-based temperature sensors. In practice, the efficiency will drop significantly as a result of the extra noise and power consumption of other blocks such as the bias circuit, ADC, and the digital back-end. It is, therefore, worthwhile to compare the obtained theoretical limits with the efficiency of actual temperature sensors.

Figure 2.11 plots the dissipated energy per conversion versus sensor resolution of various smart temperature sensors [14]. As shown, the efficiency of smart sensors published prior to the start of this research (in 2009) was limited to about $1\,nJ°C^2$, denoted by the solid FoM line. The worst-case theoretical limit of $121\,fJ°C^2$ (bias current scaling and $p = 2$) has also been projected on the same plot, facilitating the comparison. Apparently, a huge efficiency gap of about four orders of magnitude existed, thus motivating this research. Since then, the work described in this thesis (among others) has improved the energy efficiency of BJT-based temperature

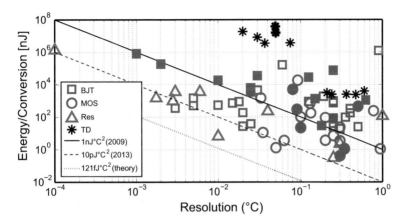

Fig. 2.11 Energy per conversion versus resolution for various smart temperature sensors [14]. The *solid and dashed FoM lines* represent the state of the art in 2009 and 2013, respectively. The *solid symbols* indicate the sensors published before 2009. The theoretical FoM line of a front-end using the (worst-case) bias current scaling ($p = 2$) is illustrated by the *dotted line*

sensors by a remarkable two orders of magnitude, represented by the dashed FoM line of 10 pJ °C^2. One of the designs achieved a resolution FoM of 11 pJ °C^2, which defined the state of the art when it was published in 2013 [15].

Recently, some more energy-efficient designs have been reported: In 2015, a resistor-based sensor was reported with a FoM of 0.65 pJ °C^2 [16]. After a single-point calibration, the sensor achieves an inaccuracy of ± 1 °C from -40 to 125 °C. In 2014, a BJT-based sensor was reported which obtains a FoM of 3.2 pJ °C^2, and an inaccuracy of ± 0.15 °C using a single-point calibration [17]. The sensor employs the concept presented in Fig. 2.5. However, its output consists of square-wave whose duty-cycle is a function of temperature. To accurately digitize this information, a counter running at several tens of MHz is required. In [17], this was done in an external FPGA, thus excluding this source of extra energy consumption from the FoM calculations.

2.4.2 Energy Efficiency Gap

As discussed, the motivation for this research was to bridge the four orders of magnitude gap between the theoretical and practical FoMs. As has been shown in the preceding, the energy dissipated by the sensor's front-end is not the limiting factor. So the limiting factor is the sensor's readout circuitry, which dissipates energy without improving the sensor's resolution. In order to minimize the energy efficiency gap, the dissipated energy as well as the excess noise introduced by readout circuitry should be minimized. Combined with the need to simultaneously achieve high resolution and accuracy, as required in most sensor applications, the result is a nontrivial task, which will be addressed in the following.

Table 2.1 Performance summary of prior-art temperature sensors published before 2009

	JSSC'98 [18]	JSSC'96 [19]	ISCAS'01 [20]	JSSC'05 [8]	JSSC'05 [1]	ISSCC'09 [2]	JSSC'09 [21]
BJTs' bias current	$36\,\mu A$	$1.1\,\mu A$	$2\,\mu A$	$2\,\mu A$	$6\,\mu A$	$1.5\,\mu A$	20–$100\,\mu A$
Total current	$1.25\,mA$	$25\,\mu A$	$100\,\mu A$	$130\,\mu A$	$75\,\mu A$	$25\,\mu A$	$1.52\,mA$
Conversion time	$30\,\mu s$	$20\,ms$	$100\,ms$	$100\,ms$	$100\,ms$	$100\,ms$	$830\,\mu s$
ADC topology	SAR	$\Delta\Sigma$-ADC	$\Delta\Sigma$-ADC	$\Delta\Sigma$-ADC	$\Delta\Sigma$-ADC	$\Delta\Sigma$-ADC	$\Delta\Sigma$-ADC
Resolution	$0.25\,°C$	$0.625\,°C$	$0.125\,°C$	$0.03\,°C$	$0.01\,°C$	$0.025\,°C$	$0.15\,°C$
FoM	$5.3\,nJ\,°C^2$	$430\,nJ\,°C^2$	$440\,nJ\,°C^2$	$32\,nJ\,°C^2$	$1.9\,nJ\,°C^2$	$3.9\,nJ\,°C^2$	$30\,nJ\,°C^2$

Table 2.1 presents an energy breakdown of several BJT-based sensors reported prior to the start of this research. The sensor presented in [18] uses $36\,\mu A$ for biasing the BJTs which is less than 3% of the total of $1.25\,mA$. It achieves a resolution of $0.25\,°C$ in a conversion time of $30\,\mu s$ and using a SAR-ADC, resulting in a FoM of $5.3\,nJ\,°C^2$. In [2], which was the most power-efficient precision sensor back in 2009, a front-end based on bias current scaling ($p = 5$) was presented. The sensor consumes $25\,\mu A$, while the current used for biasing the BJTs is only $1.5\,\mu A$: about 6% of the total. A 2nd-order $\Delta\Sigma$-ADC has been employed to achieve a resolution of $0.025\,°C$ in $100\,ms$ conversion time, thus a FoM of $3.9\,nJ\,°C^2$. A similar trend can be observed in other designs, i.e., the current used to bias BJTs is only a fraction of the total, ranging from 1.5% in [7] to 8% in [2]. This clearly indicates one major problem: the power consumed in the readout circuitry is by far dominant compared to that in the front-end.

2.4.3 ADC Topology

As can be seen from Table 2.1, $\Delta\Sigma$-ADCs have been extensively used to read out smart temperature sensors. This is because, unlike Nyquist-rate ADC topologies such as Pipeline, Flash, and Successive Approximation (SAR), $\Delta\Sigma$-ADCs have more relaxed matching requirements on the circuit elements and are capable of achieving high resolution by trading it for conversion speed. They are therefore well matched to the low speed (typically less than 10 samples/s) and high resolution (up to 15 bits) requirements of precision smart temperature sensors. In such applications, $\Delta\Sigma$-ADCs are usually operated in *incremental* mode; they are first reset at the beginning of a conversion and then operated for a fixed number of cycles [22].

Moreover, to mitigate the offset and component mismatch in the front-end, e.g., current sources and PNP transistors, which would otherwise increase the sensor's

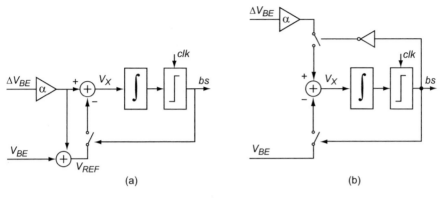

Fig. 2.12 The charge balancing in $\Delta\Sigma$ modulators is used to readout the output of the sensor front-end: (**a**) the straightforward approach, (**b**) the optimized approach

inaccuracy, dynamic error correction techniques such as chopping and dynamic-element-matching (DEM) are essential. This, in turn, results in dynamic error components in the output of the front-end, i.e., V_{BE} and ΔV_{BE}, which then need to be filtered out. The filtering characteristic of low-pass $\Delta\Sigma$ converters can be used for this purpose, thus making them a good choice for sensor applications with stringent accuracy requirements.

The use of $\Delta\Sigma$ modulators to digitize the front-end's output can be explained with the help of Fig. 2.12. As shown in Fig. 2.12a, a straightforward implementation involves generating $\alpha \cdot \Delta V_{BE}$ and V_{REF} [see Eq. (2.5)] and then digitizing their ratio with a $\Delta\Sigma$ modulator. In this example, a 1st-order modulator is formed by an integrator followed by a 1-bit quantizer in a feedback loop. Depending on the quantizer's output bitstream (bs), the integrator's input voltage V_X will be either $\alpha \cdot \Delta V_{BE}$, when bs = 0, or $-V_{BE}$, when bs = 1. The negative feedback effectively tries to balance the charges in the integrator such that the average of the integrator's output voltage tends to zero. This implies that the average value of the quantized signal bs will track the input signal and is equal to the desired μ [see Eq. (2.8)]. An equivalent charge balancing scheme can be realized with the optimized block diagram of Fig. 2.12b [5]. As shown, V_{REF} doesn't need to be explicitly generated; it can be dynamically created through the charge balancing in the loop, which simplifies the circuit implementation [1, 2].

For high energy efficiency, the modulator's output should be limited by thermal rather than quantization noise. Once the modulator is in the thermal noise limited region, its energy efficiency is theoretically no longer a function of conversion time. Compared to Nyquist-rate ADCs, however, $\Delta\Sigma$-ADCs inherently require much longer conversion times to become thermal noise limited. Although acceptable in typical temperature sensing applications, this comes at a cost of substantial energy dissipation per conversion without improving the front-end's energy efficiency, as seen from Eqs. (2.22) and (2.25). This would imply that the sensor's energy per conversion and thus FoM are substantially dominated by the readout. Using the generic approach with a 1st-order $\Delta\Sigma$-ADC requires at least 32,000 $\Delta\Sigma$-cycles to achieve a target resolution of $\pm 0.01\,°C$, equivalent to an ENOB of ≈ 15 bits. Using higher-order $\Delta\Sigma$-ADC topologies significantly reduces the conversion time, e.g., 400 and 64 $\Delta\Sigma$-cycles for the 2nd- and 3rd-order topologies, respectively. The drawback, however, is their greater complexity, reduced stable input range, larger size, and higher power consumption. By using the charge-balancing scheme of Fig. 2.12b in a 2nd-order $\Delta\Sigma$-ADC, the sensor in [2] achieves a FoM of $3.9\,nJ\,°C^2$. In recent work, this was significantly improved by using an optimized readout and the circuit flexibility offered by NPNs [23]. The sensor achieves a resolution of $0.025\,°C$ in a conversion time of 6 ms, while drawing $4.5\,\mu A$, which translates to a FoM of $24\,pJ\,°C^2$. The use of NPNs, however, requires an extra processing step, and therefore is not well suited for use in low-cost applications.

An energy-efficient alternative to a $\Delta\Sigma$-ADC could be a SAR-ADC. This is a simple structure that employs a successive approximation (binary search) algorithm in a feedback loop including a 1-bit quantizer, i.e., a comparator. The architecture operates in a "bit-at-a-time" manner, implying that an n bits conversion takes place within n steps. Compared to $\Delta\Sigma$-ADCs, this is a distinct advantage, resulting in much shorter conversions, e.g., $30\,\mu s$ for 10 bits in [18]. The simplicity in the hardware implementation of a SAR-ADC and its capacity for low-power, fast conversions has made it the most attractive choice for applications in which energy efficiency is a key requirement, e.g., biomedical and wireless sensors [24–27].

However, the resolution of SAR-ADCs depends on component matching, which means that extensive calibration is needed to achieve the minimum ENOB of 13–15 bits required by precision temperature sensors. This in turn rules out their use in low-cost applications. Moreover, the low-pass characteristic of a $\Delta\Sigma$-ADC, which is essential to the use of dynamic error correction techniques, is not provided by the SAR-ADC structure. Clearly, a trade-off exists between the high resolution/accuracy provided by $\Delta\Sigma$-ADCs and the energy efficiency of SAR-ADCs.

To summarize, the efficiency gap of sensors listed in Table 2.1 is predominantly due to their inefficient readout circuits. The employed ADCs suffer either from a long conversion time and poor power efficiency, or are not capable of providing high resolution and/or accuracy without calibration. Design of a low-power, fast, and precision ADC is therefore critical to achieving better energy efficiency in temperature sensing applications. In the next chapter, a *zoom*-ADC architecture is described which addresses this challenge.

2.5   Conclusions

In this chapter, a study of energy efficiency limits in BJT-based temperature sensors has been presented. A significant gap was observed between the energy efficiency of prior-art temperature sensors and the theoretical efficiency limits. The employed readout circuits either suffer from long conversion time and poor power efficiency, or are not capable of providing the target resolution or accuracy. To bridge this efficiency gap, a new readout architecture is clearly required. In the next chapter, a new ADC architecture will be proposed to address this issue.

References

1. M.A.P. Pertijs, K.A.A. Makinwa, J.H. Huijsing, A CMOS temperature sensor with a 3σ inaccuracy of $\pm 0.1°C$ from $-55°C$ to $125°C$. IEEE J. Solid State Circuits **40**(12), 2805–2815 (2005)
2. A.L. Aita, M.A.P. Pertijs, K.A.A. Makinwa, J.H. Huijsing, A CMOS smart temperature sensor with a batch-calibrated inaccuracy of $\pm 0.25°C$ (3σ) from $-70°C$ to $130°C$, in *Digest of Technical Papers ISSCC*, Feb 2009, pp. 342–343
3. J.F. Creemer, F. Fruett, G.C. Meijer, P.J. French, The piezojunction effect in silicon sensors and circuits and its relation to piezoresistance. IEEE Sens. J. **1**(2), 98–108 (2001)
4. F. Fruett, G.C. Meijer, *The Piezojunction Effect in Silicon Integrated Circuits and Sensors* (Kluwer Academic, Boston, 2002)
5. M.A.P. Pertijs, J.H. Huijsing, *Precision Temperature Sensors in CMOS Technology* (Springer, Dordrecht, 2006)
6. G. Wang, G.C. Meijer, The temperature characteristics of bipolar transistors fabricated in CMOS technology. Sens. Actuators A **87**, 81–89 (2000)
7. M.A.P. Pertijs, A. Niederkorn, M. Xu, B. McKillop, A. Bakker, J.H. Huijsing, A CMOS smart temperature sensor with a 3σ inaccuracy of $\pm 0.5°C$ from $-50°C$ to $120°C$. IEEE J. Solid State Circuits **40**(2), 454–461 (2005)
8. K. Souri, An energy-efficient smart temperature sensor for RFID applications. M.Sc. dissertation, Delft University of Technology, Delft, Oct. 2009
9. G.C. Meijer, Integrated circuits and components for bandgap references and temperature transducers. Ph.D. dissertation, Delft University of Technology, Delft, March 1982
10. A. Bakker, J.H. Huijsing, *High-Accuracy CMOS Smart Temperature Sensors* (Kluwer Academic, Boston, 2000)
11. M. Law, S. Lu, T. Wu, A. Bermak, P. Mak, R.P. Martins, A $1.1\mu W$ CMOS smart temperature sensor with an inaccuracy of $\pm 0.2°C$ (3σ) for clinical temperature monitoring. IEEE Sens. J. **16**(8), 2272–2281 (2016)
12. K.B. Klaassen, Digitally controlled absolute voltage division. IEEE Trans. Instrum. Meas. **24**(2), 106–112 (1975)
13. A. Hastings, *The Art of Analog Layout* (Prentice Hall, New Jersey, 2001)
14. K.A.A. Makinwa, Smart Temperature Sensor Survey [Online]. Available: http://ei.ewi.tudelft.nl/docs/TSensor_survey.xls
15. K. Souri, Y. Chae, K.A.A. Makinwa, A CMOS temperature sensor with a voltage-calibrated inaccuracy of $\pm 0.15°C$ (3σ) from $-55°C$ to $125°C$. IEEE J. Solid State Circuits **48**(1), 292–301 (2013)
16. C.-H. Weng et al., A CMOS thermistor-embedded continuous-time delta-sigma temperature sensor with a resolution FoM of $0.65pJ°C^2$. IEEE J. Solid State Circuits **50**(11), 2491–2500 (2015)
17. A. Heidary et al., A BJT-based CMOS temperature sensor with a $3.6pJ°C^2$ resolution FoM, in *Digest of Technical Papers ISSCC*, Feb 2014, pp. 224–225
18. M. Tuthill, A switched-current, switched-capacitor temperature sensor in $0.6 - \mu m$ CMOS. IEEE J. Solid State Circuits **33**(7), 1117–1122 (1998)
19. A. Bakker, J.H. Huijsing, Micropower CMOS temperature sensor with digital output. IEEE J. Solid State Circuits **31**(7), 933–937 (1996)

20. M.A.P. Pertijs, A. Bakker, J.H. Huijsing, A high-accuracy temperature sensor with second-order curvature correction and digital bus interface, in *Proceedings of ISCAS*, vol. 1, May 2001, pp. 368–371
21. H. Lakdawala et al., A 1.05V 1.6mW 0.45°C 3σ-resolution $\Delta\Sigma$-based temperature sensor with parasitic-resistance compensation in 32nm Digital CMOS process. IEEE J. Solid State Circuits **44**(12), 3621–3630 (2009)
22. J. Markus, J. Silva, G.C. Temes, Theory and applications of incremental $\Sigma\Delta$ converters. IEEE Trans. Circuits Syst. I, Fundam. Theory Appl. **51**(4), 678–690 (2004)
23. S.Z. Asl et al., A 1.55x0.85mm^2 3ppm 1.0μA 32.768kHz MEMS-based oscillator, in *Digest of Technical Papers ISSCC*, Feb 2014, pp. 226–227
24. P. Harpe et al., A 7-to-10b 0-to-4MS/s flexible SAR ADC with 6.5-to-16fJ/conversion-step, in *Digest of Technical Papers ISSCC*, Feb 2012, pp. 472–474
25. P. Harpe et al., A 0.7V 7-to-10 bit 0-to-2MS/s flexible SAR ADC for ultra low-power wireless sensor nodes, in *Proceedings of ESSCIRC*, Sept. 2012, pp. 373–376
26. N. Verma, A.P. Chandrakasan, An ultra low energy 12-bit rate-resolution scalable SAR ADC for wireless sensor nodes. IEEE J. Solid State Circuits **42**(6), 1196–1205 (2007)
27. J. Hao Cheong et al., A 400-nW 19.5-fJ/conversion-step 8-ENOB 80-kS/s SAR ADC in 0.18 − μm CMOS. IEEE Trans. Circuits Syst.-II **58**(7), 407–411 (2011)

Chapter 3
Energy-Efficient BJT Readout

3.1 Introduction

In analog circuit design, speed and precision are typically achieved at the expense of power dissipation. Therefore, the design of energy-efficient, high-resolution smart temperature sensors, as targeted in this work, is not trivial and involves fundamental trade-offs. Assuming that an ADC's power dissipation scales linearly with sampling frequency, then simply decreasing its conversion time will translate into a proportional increase in power dissipation; i.e., the ADC's energy consumption per conversion remains unchanged. Improving an ADC's energy efficiency, thus, calls for architecture-level solutions.

The generic BJT readout of Fig. 2.4 operates by digitizing ΔV_{BE} w.r.t. a stable, temperature-independent voltage reference V_{REF}. A key observation, however, is that all the necessary temperature information is present in the two voltages: V_{BE} and ΔV_{BE}. So, one could alternatively use the ratio of V_{BE} and ΔV_{BE} as a measure of temperature. In [1], a BJT-based analog front-end generates V_{BE} and ΔV_{BE} voltages at the input of a $\Delta\Sigma$-ADC. The ADC then provides the digital back-end with the nonlinear ratio $\gamma = \Delta V_{BE}/V_{BE}$. The additional signal processing, which is required to obtain the PTAT output μ as in Eq. (2.8), is carried out by the digital back-end as follows:

$$\mu = \frac{\alpha \cdot \Delta V_{BE}}{V_{BE} + \alpha \cdot \Delta V_{BE}} = \frac{\alpha \cdot \dfrac{\Delta V_{BE}}{V_{BE}}}{1 + \alpha \cdot \dfrac{\Delta V_{BE}}{V_{BE}}} = \frac{\alpha \cdot \gamma}{1 + \alpha \cdot \gamma}, \tag{3.1}$$

implying that the scaling factor α is implemented in the digital back-end. The ratio $\gamma = \Delta V_{BE}/V_{BE}$ varies between 0.019 and 0.036 over the temperature range from 45 to 175 °C. Therefore, the use of an ADC that directly digitizes γ, as presented in [1], results in a significant waste of dynamic range, much more than the 70% wasted in the generic architecture. This is demonstrated by the poor temperature resolution (0.45 °C) and FoM (324 nJ °C²) achieved in [1].

© Springer International Publishing AG 2018
K. Souri, K.A.A. Makinwa, *Energy-Efficient Smart Temperature Sensors in CMOS Technology*, Analog Circuits and Signal Processing,
DOI 10.1007/978-3-319-62307-8_3

The temperature-to-digital converter (TDC) presented in [2] switches the bias current of a diode-connected bipolar transistor from I_{bias} to $m \times I_{bias}$ in order to obtain two different base-emitter voltages: V_{BE1} and V_{BE2}. These are then digitized by an ADC. The signal processing required to determine the ratio μ is then performed in the digital back-end, e.g., the subtraction of V_{BE1} from V_{BE2} to obtain the PTAT voltage ΔV_{BE} and implementation of the scaling factor α, etc. Since ΔV_{BE} is directly calculated in the digital domain, a high resolution, high accuracy ADC is required to maintain the accuracy of the digitized ΔV_{BE} voltage. Such an ADC typically dissipates a high amount of power or requires a long conversion time, both of which would result in poor energy efficiency.

In [3], a sensor topology is presented in which a ratio $\mu' = 2 \cdot \Delta V_{BE}/(V_{BE} + 2 \cdot \Delta V_{BE})$ is digitized by a $\Delta\Sigma$-ADC. Since μ' is not a linear function of temperature, further processing in the digital back-end is performed in order to calculate the PTAT ratio as follows: $\mu_{PTAT} = 9 \cdot \mu'/(1 + 8 \cdot \mu')$. Again, the ADC's input $2 \cdot \Delta V_{BE}$ is considerably smaller than its reference voltage ($V_{BE} + 2 \cdot \Delta V_{BE}$), which translates to poor usage ($\sim$10%) of the ADC's dynamic range. The sensor requires 455 ms to achieve a resolution of 30 mK and a FoM of 4.1 nJ °C^2.

In all these examples, a nonlinear ratio between V_{BE} and ΔV_{BE} is first digitized by an ADC, while further processing is done in a digital back-end to obtain a linear function of temperature. In this approach, commonly referred to as digitally assisted analog circuit design, the high-density and low-energy per computation of digital circuits is exploited to enable circuits based on minimal-precision, low-complexity analog building blocks [4]. Since digital logic is available in all CMOS processes, this approach is attractive because it does not increase production costs. Some other potential advantages of the digitally assisted approach are scalability, lower chip area and thus lower cost, simpler architecture, and faster post-processing.

3.2 Proposed Sensor Topology

Let's consider Eq. (2.8) and reformulate it as follows:

$$\mu = \frac{\alpha \cdot \Delta V_{BE}}{V_{BE} + \alpha \cdot \Delta V_{BE}} = \frac{\alpha}{\dfrac{V_{BE}}{\Delta V_{BE}} + \alpha}. \tag{3.2}$$

By introducing a new parameter $X = V_{BE}/\Delta V_{BE}$ the equation can be simplified to:

$$\mu = \frac{\alpha}{X + \alpha}. \tag{3.3}$$

This new interpretation suggests that to obtain the PTAT ratio μ, only the ratio X needs to be digitized. A digitally implemented α and a simple division in the digital domain can then be used to compute the ratio μ.

Figure 3.1 shows the block diagram of the proposed digitally assisted bandgap temperature sensor. Compared to the block diagram of the generic bandgap temper-

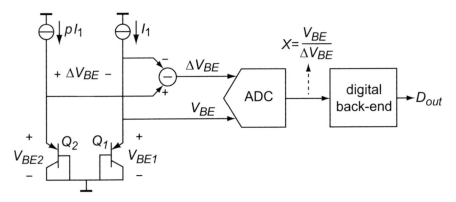

Fig. 3.1 Circuit diagram of the proposed temperature sensor based on digitizing $X = V_{BE}/\Delta V_{BE}$

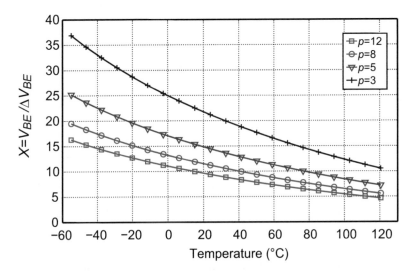

Fig. 3.2 Temperature dependence of the ratio X for different bias current ratios $p = 3, 5, 8,$ and 12

ature sensor, depicted in Fig. 2.4, the ADC only has to compute the ratio between V_{BE} and ΔV_{BE} voltages. Therefore, the two voltages no longer need to be combined to generate V_{REF}. This is one step toward simplifying the system; the temperature-independent reference voltage V_{REF} is replaced by a temperature varying voltage, i.e., V_{BE}. The ratio $X = V_{BE}/\Delta V_{BE}$, will then depend on both temperature and the collector current density ratio p used to develop ΔV_{BE}. Increasing temperature or the ratio p results in a larger PTAT voltage ΔV_{BE}, which in turn reduces the ratio X in a nonlinear fashion.

Figure 3.2 illustrates the ratio X as a function of temperature and for different bias current density ratios $p = 3, 5, 8,$ and 12. As depicted, X is a nonlinear function of temperature and decreases as p increases. A different point of view, depicted in

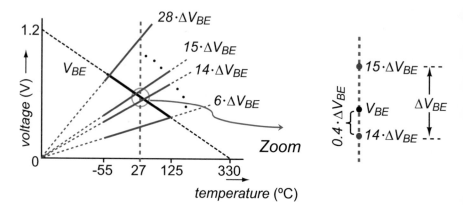

Fig. 3.3 Temperature dependence of V_{BE} and integer multiples of ΔV_{BE} from -55 to $125\,^{\circ}$C, e.g., at room temperature: $14 \cdot \Delta V_{BE} < V_{BE} < 15 \cdot \Delta V_{BE}$

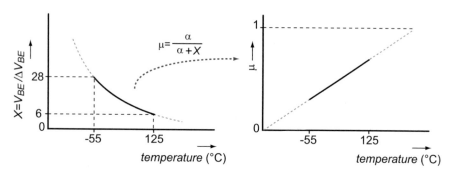

Fig. 3.4 Nonlinear $X = V_{BE}/\Delta V_{BE}(p = 5)$ and linearized $\mu = \alpha/(\alpha + X)$ as a function of temperature

Fig. 3.3, compares the base-emitter voltage V_{BE} with integer multiples of ΔV_{BE} for the case when $p = 5$. As shown, and as is also evident from Fig. 3.2, V_{BE} ranges between $6 \cdot \Delta V_{BE}$ and $28 \cdot \Delta V_{BE}$ from -55 to $125\,^{\circ}$C, and is about $14.4 \cdot \Delta V_{BE}$ at room temperature. Linearization of the ratio X can then be performed in the digital domain as described in Eq. (3.3) and graphically depicted in Fig. 3.4.

3.2.1   ADC's Resolution Requirement

For the proposed digitally assisted topology, deriving the resolution requirement on the ADC requires an analysis of the nonlinear ratio $X = V_{BE}/\Delta V_{BE}$. Considering Eq. (3.3), the derivative of ratio μ with respect to X can be described as:

$$\frac{\partial \mu}{\partial X} = \frac{-\alpha}{(X + \alpha)^2}. \tag{3.4}$$

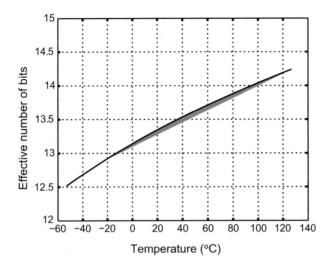

Fig. 3.5 ADC's resolution requirement versus temperature in the proposed readout approach (assuming $p = 5$) and for a target temperature resolution of $\pm 0.01\,°C$

The minimum detectable step ΔX can thus be found as:

$$\Delta\mu = \Delta X \cdot \frac{\partial\mu}{\partial X} \Rightarrow \Delta X = \frac{\Delta\mu}{\left|\dfrac{\partial\mu}{\partial X}\right|}, \tag{3.5}$$

implying a temperature-dependent resolution due to the temperature dependence of the derivative $\partial\mu/\partial X$. Moreover, for a given temperature resolution, the required resolution will depend on the coefficient α, and thus on the current density ratio p, as is evident from Eq. (2.7).

The ADC's resolution can be calculated as follows:

$$\text{ENOB} \approx \log_2\left(\frac{X_{\text{FS}}}{\Delta X}\right) - 1, \tag{3.6}$$

where X_{FS} is the full scale, i.e., $X_{\text{FS}} = X_{\max} - X_{\min} = 22$, for $p = 5$. For the target sensor resolution of $\pm 0.01\,°C$, the ADC's required resolution as a function of temperature is shown in Fig. 3.5. As shown, the resolution requirement is slightly temperature dependent, with a maximum value of 14.3 bits at 125 °C, which is about 0.5 bits less than that of the generic approach. The corresponding impact on conversion time and power consumption will depend on the chosen ADC architecture, which is the subject of the following discussion.

3.3   The Zoom-ADC: An Energy-Efficient ADC

3.3.1   Introduction

In general, temperature changes are rather slow. Therefore, the ratio X can be accurately digitized by a two-step ADC, in which a full-range conversion first obtains a coarse estimate of the input level. This is followed by a low-range, but high resolution, fine conversion to obtain an accurate estimate of the input level. In the proposed zoom-ADC, the strengths of both SAR-ADC and $\Delta\Sigma$ converters are combined into a two-step conversion scheme, as will be explained in the following.

3.3.2   Topology

In the proposed sensor topology, the input to the ADC, $X = V_{\mathrm{BE}}/\Delta V_{\mathrm{BE}}$, is always greater than one, and thus can be expressed in terms of integer and fractional parts, as follows:

$$X = \frac{V_{\mathrm{BE}}}{\Delta V_{\mathrm{BE}}} = n + \mu', \tag{3.7}$$

where n is the integer part of X, and μ' is its fractional part, i.e., $0 < \mu' < 1$. As we observed in Fig. 3.2, the magnitude of X, and thus the integer value n, depends, apart from temperature, on the collector current density ratio p used to generate ΔV_{BE}. In the rest of this chapter we assume $p = 5$. This implies that X ranges from 6 to 28 over the military temperature range, i.e., from -55 to $125\,^{\circ}\mathrm{C}$.

To determine X, the two parameters n and μ' are determined in a two-step manner, as shown in Fig. 3.6; In the first step the integer n is found in a coarse conversion phase (Fig. 3.6a). Knowing n, the fraction μ' can then be readily resolved in the succeeding fine conversion step by zooming into the range from n to $n+1$, as graphically depicted in Fig. 3.6b. Since the full scale range of the fine conversion is now quite small, the resolution requirements of the fine converter are greatly relaxed, thus leading to a simple implementation and short conversion times, concurrently. Moreover, in contrast to the previously described readout architectures, the dynamic range of the zoom-ADC is now fully utilized, thus enabling energy-efficient conversions.

3.3.3   Coarse Converter

In order to determine the integer value n, a low-resolution converter is sufficient. As shown in Fig. 3.2, for practical p ratios a resolution of about 5–7 bits in the coarse converter is sufficient to find the integer n. As explained before, this can be quickly done by means of a SAR-ADC. This combines a simple hardware implementation

Fig. 3.6 Temperature
dependence of $X = n + \mu'$
from -55 to $125\,^{\circ}$C. The
integer n ranges between 6
and 28 (**a**), while the fraction
μ' ranges between 0 and 1 (**b**)

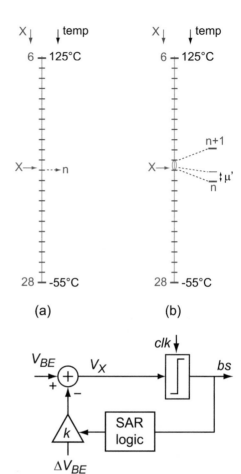

(a)   (b)

Fig. 3.7 Block diagram of
the zoom-ADC during the
coarse conversion

with low-power operation, both of which are key requirements for our target
application. Assuming $p = 5$ implies $6 \leq n \leq 28$, and thus a 5-bit SAR-ADC.

Figure 3.7 shows the proposed block diagram of the coarse converter. In a
feedback loop, the base-emitter voltage V_{BE} is compared to integer multiples of
ΔV_{BE}, i.e., $k \cdot \Delta V_{\mathrm{BE}}$, where $k = 1, 2, \ldots, 28$. A clocked comparator then performs
the comparison by simply detecting the sign of:

$$V_X = V_{\mathrm{BE}} - k \cdot \Delta V_{\mathrm{BE}}. \tag{3.8}$$

The comparison result bs is then used by the SAR-logic, which, in turn, properly
updates the multiplication factor k for the next comparison step. The procedure will
be continued until the SAR-logic finds the region in which:

$$n \cdot \Delta V_{\mathrm{BE}} < V_{\mathrm{BE}} < (n + 1) \cdot \Delta V_{\mathrm{BE}}. \tag{3.9}$$

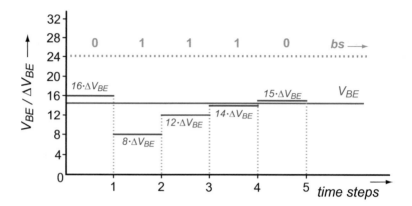

Fig. 3.8 The successive approximation steps during the coarse conversion. The unknown n is determined within five approximation steps

At room temperature, this condition is satisfied for $n = 14$ (see Fig. 3.3). Since $n \leq 28$, the SAR-logic only needs five steps to find n. Figure 3.8 shows the comparison steps performed by SAR-logic to find n at room temperature.

An alternative approach to further simplify the coarse conversion would be to replace the SAR-logic with a linear search algorithm as follows: the control logic sets $k = 6$, which corresponds to the high end of temperature range, i.e., 125 °C. Since $V_{BE} < 6 \cdot \Delta V_{BE}$ over the temperature range from −55 to 125 °C, the comparison result will be always 1. The control logic then starts ramping up the comparison level by 1, i.e., $k = 7, 8, \ldots$ until it detects a zero at the comparator's output implying that $V_{BE} < (n + 1) \cdot \Delta V_{BE}$. At room temperature, ten comparison steps are required, while at −55 °C the maximum number of 23 steps are required. Although this approach requires simpler control logic, the price to pay is a longer (and temperature dependent) conversion time in the coarse phase, e.g., 1–23 versus five comparison steps for the case of a SAR-logic.

3.3.4   Fine Converter

Having found the integer n, the next step is to determine the fraction μ'. By rewriting Eq. (3.3), this can be expressed as follows:

$$\mu' = \frac{V_{BE} - n \cdot \Delta V_{BE}}{\Delta V_{BE}}. \tag{3.10}$$

In other words, the fraction μ' can be found by resolving the *residue* ($V_{BE} - n \cdot \Delta V_{BE}$) over the range ΔV_{BE}. This can be done by using a $\Delta\Sigma$-ADC, whose input and reference voltages are set to ($V_{BE} - n \cdot \Delta V_{BE}$) and ΔV_{BE}, respectively, as shown in Fig. 3.9. It consists of a summation node, an integrator, a 1-bit quantizer (a clocked comparator), and a 1-bit digital-to-analog converter (DAC). The sign

Fig. 3.9 Block diagram of the zoom-ADC during the fine conversion

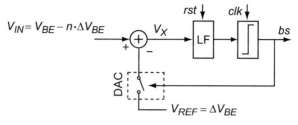

Fig. 3.10 An equivalent block diagram to perform the fine conversion

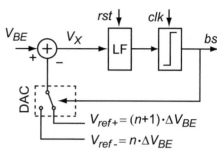

of the integrator's output, V_{INT}, is determined on the rising edges of the clock. The comparison result bs is then fed back via the DAC, which determines whether ΔV_{BE} has to be subtracted from the input $V_{IN} = V_{BE} - n \cdot \Delta V_{BE}$ or not. Therefore, the integrator's input V_X can be expressed as:

$$V_X = \begin{cases} V_{BE} - n \cdot \Delta V_{BE} & \text{if bs} = 0 \\ V_{BE} - (n+1) \cdot \Delta V_{BE} & \text{if bs} = 1. \end{cases} \qquad (3.11)$$

A loop filter (LF) then processes the difference between the input voltage V_{IN} and the feedback voltage. Since the net integrated charge is forced to be approximately zero by the feedback loop, the bitstream average is the desired fraction $\mu' = (V_{BE} - n \cdot \Delta V_{BE})/\Delta V_{BE}$. In practice, a decimation filter processes bs to calculate μ'. Depending on the order of loop filter, different types of decimation filters could be used to optimally decimate the produced bitstream bs. A discussion on the different types of decimation filters and their pros and cons is, however, out of the scope of this thesis and can be found in [5].

A more elegant way to find the fraction μ' is shown in the block diagram of Fig. 3.10. This topology is equivalent to the one in Fig. 3.9, but results in a much simpler circuit-level implementation. Depending on the bitstream value bs, either $n \cdot \Delta V_{BE}$ or $(n + 1) \cdot \Delta V_{BE}$ will be subtracted from the input V_{BE}, thus simply implementing Eq. (3.10). Likewise, $(V_{BE} - n \cdot \Delta V_{BE})$ is integrated when bs = 0. When bs = 1 the input to the integrator V_X will be reduced by ΔV_{BE}, i.e., $V_X = V_{BE} - (n + 1) \cdot \Delta V_{BE}$, and thereby reduces the accumulated value. Similarly the charge balancing forces the average charge accumulated in the integrator to be (approximately) zero, and hence Eq. (3.10) holds.

Comparing the block diagrams of the coarse and fine converters shown in Figs. 3.10 and 3.7 reveals the similarities between the two converters; they both sample $(V_{BE} - k \cdot \Delta V_{BE})$, and use a 1-bit quantizer, i.e., a comparator. Sharing hardware between the two converters enables a compact implementation, as will be discussed in the following chapters.

A key observation is that the range of the $\Delta\Sigma$-ADC employed in the fine conversion is reduced to ΔV_{BE}, which is $\approx 40\,$mV at room temperature ($p = 5$). Compared to the bandgap voltage reference of $\approx 1.2\,$V used in conventional approaches, this results in significantly shorter and thus more energy-efficient conversions. As a result of the reduced full scale range in the fine step, the error signals processed by the loop filter have small amplitudes, which can be readily handled by low-power, low-swing integrators. Furthermore, since the fine conversion step incorporates a $\Delta\Sigma$-ADC, it leverages the high resolution capability of $\Delta\Sigma$-ADCs without relying on matching of circuit elements. Moreover, dynamic techniques such as chopping, auto-zeroing (AZ), and dynamic element matching (DEM) [5, 6] can still be used in the zoom-ADC, to achieve the precision required in temperature sensors. In the following, a more detailed system-level analysis of the proposed zoom-ADC is presented.

3.3.5 System-Level Considerations

In this section the system-level design of a zoom-ADC is addressed, together with some considerations for its practical implementation.

3.3.5.1 Redundancy and Guard-Banding

In practice, the quantization errors of the coarse converter, which are caused by the comparator's input-referred noise and offset as well as by mismatch in the feedback DAC, could result in incorrect n values, as shown in Fig. 3.11. This is especially problematic when $X \approx n$ or $(n + 1)$. As shown, the coarse result could then be either $(n - 1)$ or $(n+1)$, when $X \approx n$ or $X \approx (n+1)$, respectively. The fine conversion will then zoom into the wrong region, resulting in clipping or out-of-ranging, both of which would result in significant errors at the ADC's output, up to 1-LSB of the coarse conversion.

To avoid such out-of-range errors, redundancy, which is often employed in two-step ADCs [7, 8], is also employed in the zoom-ADC to accommodate the SAR-ADC's quantization errors and relax its required accuracy. This is done by extending the fine conversion range and making it equal to 2-LSBs of the coarse conversion, i.e., $2 \cdot \Delta V_{BE}$. This redundancy ensures that the input X will always lie safely within the input range of the fine $\Delta\Sigma$-ADC. Since extending the range comes at the price of reduced resolution in the $\Delta\Sigma$ phase, the number of cycles should be appropriately increased to maintain the target ENOB.

Fig. 3.11 Accuracy in coarse step: Ideal (*left*) and practical (*right*) situations

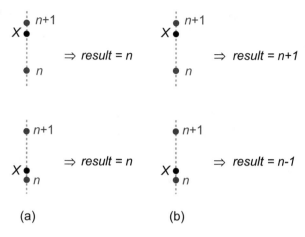

(a)   (b)

Fig. 3.12 Guard-banding mechanism ensures that X is always roughly in the middle of the extended range, thus avoiding out-of-ranging

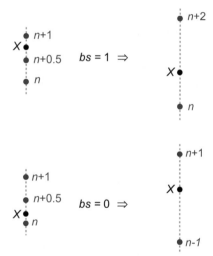

The overlap between the two conversion steps is readily implemented by setting the feedback reference range of the $\Delta\Sigma$ modulator to $2 \cdot \Delta V_{BE}$. However, ensuring that the input X is always within the extended range requires an extra comparison step. As shown in Fig. 3.12, this involves comparing V_{BE} to $(n+0.5)\cdot\Delta V_{BE}$ at the end of the coarse conversion, and is referred to as the "guard-banding" step throughout this thesis. The total number of coarse comparison steps, therefore, increases by one. Depending on the result, the references of the fine converter can then be set to cover the range from n to $(n + 2)$ or from $(n - 1)$ to $(n + 1)$, for bs = 1 and bs = 0, respectively. This can be implemented by modifying the block diagram of Fig. 3.10 so as to set the feedback references to $(n - 1)\cdot\Delta V_{BE}$ and $(n + 1)\cdot\Delta V_{BE}$ or to $n \cdot \Delta V_{BE}$ and $(n + 2)\cdot\Delta V_{BE}$. Assuming a 5-bit coarse conversion, and a fine input range of $2 \cdot \Delta V_{BE}$, the error in the coarse ADC should be less than $\pm 0.5 \cdot \Delta V_{BE}$, which corresponds to a worst case error of $\pm 15\,\text{mV}$ at $-55\,°\text{C}$.

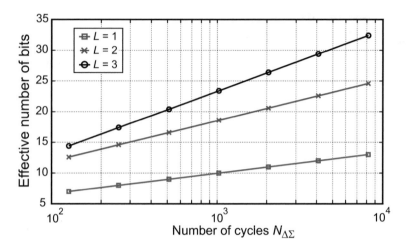

Fig. 3.13 ENOB versus number of cycles $N_{\Delta\Sigma}$ for conventional incremental $\Delta\Sigma$-ADCs. L is the loop filter's order

By combining the results from the two conversion steps, the digital output X_{out} of the zoom-ADC can then be expressed as follows:

$$X_{out} = n' + 2 \cdot \mu'', \tag{3.12}$$

where n' and μ'' are the results of the coarse and fine steps after guard-banding step, i.e., n' is equal to either $(n-1)$ or n, while $0 < \mu'' < 1$ is the resolved residue over the extended range of $2 \cdot \Delta V_{BE}$, and thus should be scaled before adding to n'.

3.3.5.2   Number of Cycles

When operated in incremental mode, the achievable ENOB of a $\Delta\Sigma$ modulator depends on three parameters: the order of the loop filter, the quantizer resolution, and the number of clock cycles $N_{\Delta\Sigma}$. Due to its ideal linearity, a 1-bit quantizer is best suited for precision applications. It is also power and area efficient, since it only requires a single comparator. To illustrate the modulator's design space, Fig. 3.13 shows the simulated ENOB of a 1-bit incremental $\Delta\Sigma$ modulator versus the number of cycles $N_{\Delta\Sigma}$ for various modulator orders (L). It can be seen that for the same ENOB, $N_{\Delta\Sigma}$ can be reduced by using higher order loop filters. However, this is at the expense of more complexity and power consumption, as well as a reduction in the range of input signals for which the modulator is stable. For example, the usable input range of a 3rd order $\Delta\Sigma$ modulator is only about 67% of its reference range [6].

The maximum ENOB obtainable from the 1st- and 2nd-order zoom-ADCs versus $N_{\Delta\Sigma}$ at 25 °C is illustrated in Fig. 3.14, assuming a coarse ADC resolution of 5 bits. In order to achieve the target ENOB of 13.4 bits (to allow a temperature

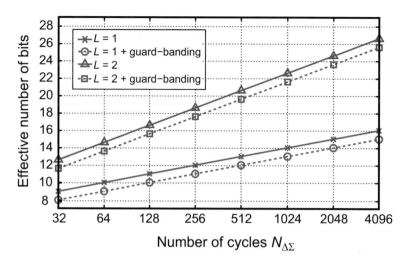

Fig. 3.14 ENOB versus number of cycles $N_{\Delta\Sigma}$ (at 25 °C) for the 1st and 2nd-order zoom-ADC employing a 5-bit coarse converter

resolution of ±0.01 °C at 25 °C, as in Fig. 3.5), a 1st-order zoom-ADC would theoretically require ≈680 cycles. The use of a 2nd-order zoom-ADC reduces the required number of cycles to ≈43. Considering the increased range due to guard-banding, these numbers will increase to about 1360 and 60 for the 1st- and 2nd-order zoom-ADCs, respectively. However, for high energy efficiency, the modulator's output should be limited by thermal rather than quantization noise. This translates to a slight increase in the number of cycles shown in Fig. 3.14. Moreover, the full scale range of the fine converter is temperature dependent $(2 \cdot \Delta V_{BE})$, as is its quantization error. On the other hand, due to the nonlinear ENOB requirements versus temperature, higher resolution is required at higher temperatures (see Fig. 3.5). This again translates into an increase in the number of cycles in order to meet the target resolution over the whole temperature range.

3.3.5.3 Signal Swing

The loop filter of a $\Delta\Sigma$ modulator consists of a number of integrator blocks, each of which will be implemented by an amplifier. The first integrator processes the error signal, i.e., the difference between the input and reference signals, whose amplitudes define the integrator's output swing. In practice, the output swing defines the amplifier's topology, and eventually its power efficiency. In conventional single-bit $\Delta\Sigma$ modulators, the output of the first integrator always includes a large input-signal related component, which results in poor power efficiency. In the proposed zoom-ADC, this issue is mitigated because the error signal processed by the loop filter of the fine $\Delta\Sigma$-ADC is proportional to the quantization error of the coarse conversion step, and is thus significantly smaller.

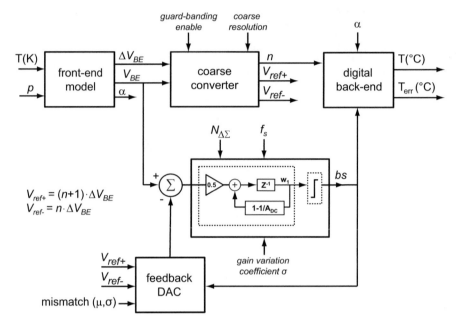

Fig. 3.15 Simplified block diagram of the behavioral model of the zoom-ADC

Fig. 3.16 Maximum signal swing of the fine $\Delta\Sigma$-ADC's first integrator as a function of the coarse step resolution (a voltage gain of 0.5× was assumed)

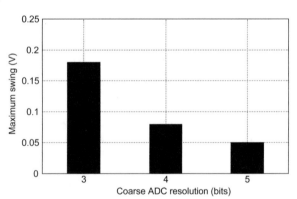

To verify this, a behavioral model for a 1st-order zoom-ADC was developed. Figure 3.15 shows a simplified block diagram of this model, in which various design parameters such as the current ratio p, the resolution of the coarse converter, the mismatch parameters of the feedback loop, the DC gain of the integrator, etc. can be adjusted. Using this model, Fig. 3.16 shows the maximum swing of the integrator's output as a function of the coarse ADC's resolution at 25 °C. As shown, the output swing is inversely proportional to the resolution of the coarse ADC. For a 5-bit coarse ADC and assuming a voltage gain of 0.5×, an output swing of only ±50 mV is required. Clearly, guard-banding effectively reduces the resolution of coarse converter by 1 bits, resulting in larger error signals and slightly increasing the output swing.

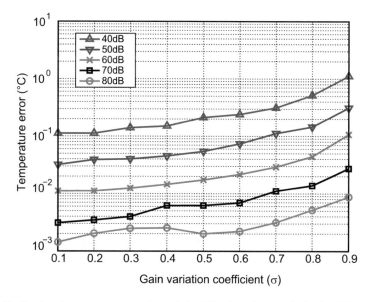

Fig. 3.17 Simulated temperature error due to finite DC gain and gain variation in the integrator of a 1st-order zoom-ADC with 5-bit coarse conversion

A similar model can be developed for the conventional charge balancing configuration (see Fig. 2.12) [9, 10]. The result is that the first integrator has to handle an output swing of about ±300 mV for a similar voltage gain of 0.5×. Apparently, zooming significantly relaxes the opamp's output swing and settling requirements, and thus enables a simple and power-efficient realization, as will be shown in the following chapters.

3.3.5.4 Integrator Gain

The finite DC gain and gain variation of the integrators in the loop filter are critical error sources. Due to the finite DC gain of the amplifier used in the integrator, the charge on the integration capacitors will leak away. The impact caused by such a leaky integrator is nonlinearity, which reduces the achievable effective number of bits (ENOB). An amplifier's DC gain can usually be approximated by a third-order polynomial [11] as follows:

$$A_{DC}(V_{out}) \approx A_{DC} \cdot \left(1 - \sigma \left| \frac{V_{out}}{V_{max}} \right|^3 \right), \tag{3.13}$$

where A_{DC} is the DC gain at the mid-level output, σ is the gain variation coefficient, V_{out} is the output swing, and V_{max} is the maximum output swing. The model in Fig. 3.15 has been used to verify the impact of such variations on the zoom-ADC's performance. Figure 3.17 shows the simulated error in degrees Celsius due to the finite DC gain and gain variation of the integrator of a 1st-order zoom-ADC, assuming a 5-bit coarse conversion. In this plot, the impact of gain variations on the coarse converter is neglected, since in this step the integrator only acts as a sample-and-hold amplifier for the quantizer, and thus for practical DC gain values, its gain variations will not impact the coarse conversion result. As shown, a DC gain of 80 dB is required to ensure a temperature error below ± 10 mK, assuming $\sigma = 0.9$, which represents a significant gain variation coefficient. In order to achieve similar performance with the conventional charge balancing scheme used in [9, 10], a minimum DC gain of 100 dB would be required, which is considerably higher and more difficult to achieve, especially at low supply voltages in modern processes, e.g., 1.8 V in the target 160 nm CMOS and sub-1 V in nano-scale processes.

Given the reduced output signal swing, the OTA of a zoom-ADC does not slew and so its settling follows a single pole response. In this case, incomplete settling only results in a fixed gain error in the integrator. As long as this error does not significantly alter the loop filter's transfer function, the modulator's performance will not be impaired.

3.3.5.5   DAC Mismatch

As shown in Fig. 3.7, the proposed zoom-ADC's $\Delta\Sigma$ modulator uses a 1-bit quantizer, and so its linearity is not limited by quantizer offset and offset spread. However, the ADC's overall linearity is limited by the nonlinearity of its multi-bit DAC, which is caused by the mismatch of the various DAC elements, e.g., the mismatch between the unit capacitors of a unary-weighted capacitor DAC (CDAC). As in other two-step ADC structures, this mismatch could result in ADC nonlinearity. The standard metric to evaluate this is to measure the ADC's integral nonlinearity (INL) and differential nonlinearity (DNL). While INL is considered as the error with respect to a best linear-fit, DNL reflects the discontinuities in the ADC's transfer function.

Although the resulting nonlinearity can be improved by calibration and/or trimming of the DAC references, the associated calibration time significantly increases production costs, thus prohibiting the use of such techniques for large volume production. A suitable solution would be to apply DEM to the unit elements of the multi-bit DAC. During the fine conversion phase, a subset of DAC elements which form the $\Delta\Sigma$-ADC's references can be dynamically interchanged, thus averaging out their mismatch. The simulation model of Fig. 3.15 can be used to investigate the effect of the mismatch between the unit elements of the feedback DAC. Figure 3.18, top, shows the simulated INL or temperature error of a 1st-order zoom-ADC with the random mismatch of the feedback DAC, assuming a normal distribution with mean = 0 and $\sigma = 1\%$. As shown, the resulting temperature errors can be as high as $\pm 0.2\,^{\circ}$C, which is unacceptably large for precision applications. Applying DEM will improve the accuracy to $\pm 0.02\,^{\circ}$C, representing a 10$\times$ improvement as shown in Fig. 3.18, bottom.

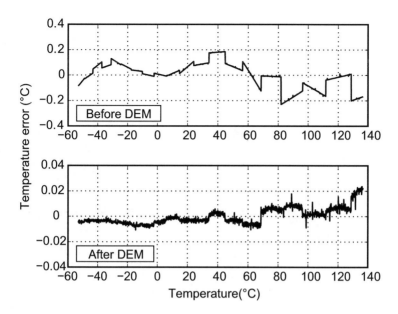

Fig. 3.18 Simulated temperature error in the 1st-order zoom-ADC, assuming a mismatch with normal distribution (zero mean and $\sigma = 1\%$) between the DAC elements: before (*top*) and after (*bottom*) applying DEM ($N_{\Delta\Sigma} = 1024$)

Measuring the DNL of a smart temperature sensor is not trivial, as it requires the generation of temperature steps that are smaller than the ADC's quantization error. A more practical alternative is to take the difference between consecutive ADC outputs, thereby defining a pseudo-DNL function. If the rate of change at the input of the ADC is less than its conversion speed, any discontinuities in the ADC's characteristics will be captured by this experiment, which can be readily performed by exposing the sensor to a thermal ramp. The simulated pseudo-DNL for a normal distribution (mean = 0 and $\sigma = 1\%$) between the DAC's unit elements is shown in Fig. 3.19, top. As shown, a significant discontinuities between the different fine segments is observed, e.g., $\pm 0.2\,^\circ$C at $X = 7.5$ and $X = 10.5$. Applying DEM reduces these to well below $\pm 0.015\,^\circ$C. In practice, careful design and layout can reduce the mismatch of the DAC's unit elements to less than 1%.

3.4 Curve Fitting and Trimming

So far, the design of an ADC to digitize the ratio X has been discussed. However, further digital signal processing is required before an output D_{out} in degrees Celsius can be obtained. By combining Eqs. (3.3) and (2.9), D_{out} can be expressed as follows:

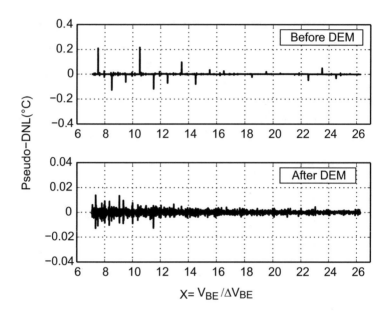

Fig. 3.19 Simulated pseudo-DNL of the 1st-order zoom-ADC, assuming a mismatch with normal distribution (zero mean and $\sigma = 1\%$) between the DAC elements: before (*top*) and after (*bottom*) applying DEM ($N_{\Delta\Sigma} = 1024$)

$$D_{\text{out}} = A \cdot \frac{\alpha}{X + \alpha} + B, \tag{3.14}$$

indicating that the parameters α, A, and B should be determined before D_{out} can be obtained. The mapping parameter α can be found by solving Eq. (2.7), which requires the knowledge of the design parameters p and $S_{V_{\text{BE}}}^T$. In practice, during a *calibration* phase the ADC's output X is carefully characterized over the temperature range of interest. During this phase, a *reference* sensor, in thermal equilibrium with the device under test (DUT), provides accurate temperature information T_{ref}, which is logged along with X [5]. The optimum mapping parameter α_{opt} can then be found by applying a least-squares linear fit to the resulting μ from Eq. (3.3). For a properly designed sensor front-end and readout circuit, the resulting α_{opt} should match that obtained from Eq. (2.7). Assuming that $p = 5$, and $S_{V_{\text{BE}}}^T = -2\,\text{mV/°C}$, a value of $\alpha_{\text{opt}} = 15.218$ is obtained by both methods.

By means of linear fitting of μ to the temperature information T_{ref}, obtained during the calibration phase, the scaling parameters A and B can also be found. It can be seen from Eq. (3.14) that their values will also depend on α. This becomes more clear by applying a Taylor expansion of Eq. (3.14) around α_{opt} as follows:

$$D_{\text{out}}(\alpha_{\text{opt}}) = A \cdot \frac{\alpha_{\text{opt}} \cdot \Delta V_{\text{BE}}}{V_{\text{BE}} + \alpha_{\text{opt}} \cdot \Delta V_{\text{BE}}} + B + A \cdot \frac{\Delta V_{\text{BE}} \cdot V_{\text{BE}}}{(V_{\text{BE}} + \alpha_{\text{opt}} \cdot \Delta V_{\text{BE}})^2} \cdot (\alpha - \alpha_{\text{opt}}) + \cdots, \tag{3.15}$$

Fig. 3.20 The required adjustments of the gain and offset parameters A, B, as a result of $\pm 10\%$ deviation of α from α_{opt}

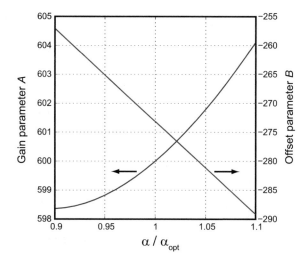

indicating that at a given temperature, any deviation of α from α_{opt} will require the adjustment of both A and B parameters, in order to achieve the same D_{out}. Figure 3.20 shows the resulting modifications in A and B, as a function of $\alpha = \alpha_{\text{opt}} \pm 10\%$. As can be seen, B linearly decreases with α, while A increases parabolically. Given that for $\alpha = \alpha_{\text{opt}}$, $A = 600$ and $B = -273.14$, a 1% increase from α_{opt} requires A and B to change by 0.052% and 0.59%, respectively, indicating a significantly larger sensitivity of B to variations in α.

So far, V_{BE} is assumed to be linear with a sensitivity $S_{V_{\text{BE}}}^{T} = -2\,\text{mV/°C}$, which is not an accurate assumption. In practice, V_{BE} exhibits nonlinearity or *curvature* in the order of few mV, depending on the bias conditions, e.g., using a PTAT bias current for a substrate PNP results in a V_{BE} curvature up to $\approx 3\,\text{mV}$ from -55 to $125\,°\text{C}$ [5]. By applying the values of α_{opt}, A, and B, as obtained so far, a systematic nonlinearity error up to $\approx 1\,°\text{C}$ at the sensor's output can be measured. Although this can be compensated by digital post-processing, this approach increases the complexity of the digital back-end. As shown in [12, 13], a more straightforward compensation can be achieved by slightly increasing the value of α_{opt}, in order to minimize the nonlinearity of μ due to the curvature in V_{BE}. The value of parameters A and B should then be slightly adjusted to obtain the required linear mapping between μ and the output D_{out}. The resulting residual nonlinearity is then reduced to less than $\pm 0.1\,°\text{C}$ from -55 to $125\,°\text{C}$ [12], which is sufficiently small in most of the applications.

In practice, the sensor's output deviates from the desired value due to the process variation of different on-chip components. During the calibration phase, the resulting error can be accurately measured and then nulled by adjusting one of the different possible *trimming* knobs. In a BJT-based sensor, the use of dynamic correction techniques ensures that the dominant source of inaccuracy is the process spread of V_{BE}, which is PTAT in nature, and therefore can be trimmed at a single temperature [5]. Assuming $\pm 2\,\text{mV}$ spread of V_{BE} (at room temperature), Fig. 3.21

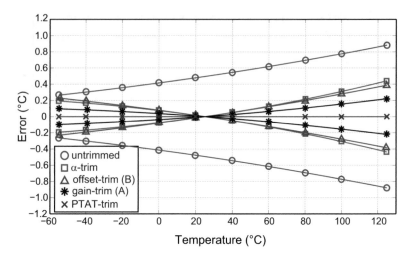

Fig. 3.21 Error due to the PTAT spread in V_{BE} ($\pm 2\,\text{mV}$ at $25\,^\circ\text{C}$), before and after trimming: different single-point trim methods are employed

shows that the resulting PTAT error over the temperature range from -55 to $125\,^\circ\text{C}$ is about $\pm 0.9\,^\circ\text{C}$ at $125\,^\circ\text{C}$. The most effective way to trim such an error is to apply a PTAT compensation to V_{BE}. A straightforward way of doing this is to trim the BJT's bias current [5]. Alternatively, the adjustment can be made in the digital domain, by modifying Eq. (2.9) as follows:

$$D_{\text{out}} = A \cdot \frac{\mu}{1 - \gamma_D \mu} - B. \tag{3.16}$$

Here γ_D is a calibration constant that is determined as follows:

$$\gamma_D = \frac{1}{\mu} - \frac{1}{\mu_{\text{ideal}}}, \tag{3.17}$$

where μ_{ideal} is the desired ratio at the calibration temperature [5]. The implementation of digital PTAT trim, therefore, requires extra digital signal processing, including a division and a multiplication, which translates to extra chip area.

However, Eq. (3.14) suggests that the parameters A, B, and α can also be used as trimming knobs. For example, an offset trim can be readily implemented by adjusting B. This choice, however, is not optimal as it does not completely cancel out the error due to the V_{BE}'s PTAT spread. For the $\pm 2\,\text{mV}$ V_{BE} spread, an offset trim at room temperature reduces the inaccuracy to around $\pm 0.4\,^\circ\text{C}$ at $125\,^\circ\text{C}$, as shown in Fig. 3.21. Depending on the target accuracy, offset trim can be considered as an attractive alternative, due to its simplicity of implementation. In the proposed readout architecture in this work, α is implemented in the digital domain, and thus can also be used as a trim knob. As shown in Fig. 3.21, a single-point α-trim at room temperature results in a slightly greater inaccuracy compared to that obtained with an offset trim; around $\pm 0.45\,°C$ at $125\,°C$. Since the PTAT error is a particular sort of gain error, a higher accuracy is expected by trimming A. As shown, this reduces the inaccuracy to about $\pm 0.25\,°C$ at $125\,°C$.

The discussion above suggests that PTAT trim is the most suitable trimming method for BJT-based sensors, as it fully compensates for the PTAT spread of V_{BE}. In practice, however, any residual non-PTAT error will impact the effectiveness of trimming, and therefore a complete cancellation will not be possible. Various error sources can contribute to such non-PTAT residuals, e.g., the high-order nonlinearity or curvature of V_{BE}, packaging stress, and other residual errors in the sensor front-end and readout circuit [5]. Moreover, any error during the calibration process will also directly impact the trimming, and therefore the sensor's ultimate accuracy. The adopted trimming method, therefore, depends on the nature of any non-PTAT residual errors in the design, the target accuracy, the hardware constraints on the choice of trimming parameter, and the calibration accuracy.

3.5 Conclusions

In this chapter, a new readout architecture was proposed to bridge the energy efficiency gap of the BJT-based temperature sensors. Given the fact that temperature is a slowly changing quantity, a two-step zoom-ADC architecture was proposed, which combines the speed of a coarse SAR-ADC, with the high resolution/accuracy of a fine $\Delta\Sigma$-ADC. In the zoom-ADC, the full-scale range of the fine converter is considerably reduced, thus notably relaxing various key requirements such as the number of $\Delta\Sigma$-cycles and the DC gain and swing of the loop filter. Since both conversion time and power efficiency can be improved in this architecture, a substantial energy efficiency improvement can be achieved with the zoom-ADC. Lastly, the fact that dynamic correction techniques can be used in the fine conversion phase ensures that the accuracy of the zoom-ADC can be as good as that of conventional $\Delta\Sigma$-ADC architectures.

The required curve fitting to achieve the digital output in degrees Celsius was also discussed. This results in a set of parameters A, B, and α which determine the appropriate mapping. As shown, in order to minimize the nonlinearity of the sensor's output, the value of α should be optimally set. It was also observed that any deviation from this optimal value will call for small adjustments of the gain and offset parameters A and B. Finally, different trimming methods were explored to compensate for the sensor's dominant source of inaccuracy: the PTAT spread of V_{BE}. As discussed, a PTAT trim is the most suitable trimming method to cancel out such error. The other fitting parameters such as A, B, and α can also be used as trimming knobs, but result in significantly greater inaccuracy when compared to PTAT trim.

References

1. H. Lakdawala et al., A 1.05 V 1.6 mW 0.45 °C 3σ-resolution $\Delta\Sigma$-based temperature sensor with parasitic-resistance compensation in 32 nm digital CMOS process. IEEE J. Solid-State Circuits **44**(12), 3621–3630 (2009)
2. Lin et al., CMOS temperature-to-digital conversion with digital correction, U.S. Patent US8167485 B2, April 2008
3. F. Sebastiano et al., A 1.2-V $10 - \mu$W NPN-based temperature sensor in 65-nm CMOS with an inaccuracy of ±0.2 °C (3σ) from -70 °C to 125 °C. IEEE J. Solid-State Circuits **45**(99), 2591–2601 (2010)
4. B. Murmann, Digitally assisted analog circuits, in *Proceedings of the IEEE Micro*, vol. 26(2), March–April 2006, pp. 38–47
5. M.A.P. Pertijs, J.H. Huijsing, *Precision Temperature Sensors in CMOS Technology* (Springer, Amsterdam, 2006)
6. J. Markus, J. Silva, G.C. Temes, Theory and applications of incremental $\Sigma\Delta$ converters. IEEE Trans. Circuits Syst. I: Fundam. Theory Appl. **51**(4), 678–690 (2004)
7. P. Harpe et al., A 0.7 V 7-to-10 bit 0-to-2 MS/s flexible SAR ADC for ultra low-power wireless sensor nodes, in *Proceedings of the ESSCIRC*, Sept. 2012, pp. 373–376
8. N. Verma, A.P. Chandrakasan, An ultra low energy 12-bit rate-resolution scalable SAR ADC for wireless sensor nodes. IEEE J. Solid-State Circuits **42**(6), 1196–1205 (2007)
9. M.A.P. Pertijs, K.A.A. Makinwa, J.H. Huijsing, A CMOS temperature sensor with a 3σ inaccuracy of ±0.1 °C from -55 °C to 125 °C. IEEE J. Solid-State Circuits **40**(12), 2805–2815 (2005)
10. A.L. Aita, M.A.P. Pertijs, K.A.A. Makinwa, J.H. Huijsing, A CMOS smart temperature sensor with a batch-calibrated inaccuracy of ±0.25 °C (3σ) from -70 °C to 130 °C, in *Digest of Technical Papers ISSCC*, Feb 2009, pp. 342–343
11. H. Park, K. Nam, D.K. Su, K. Vleugels, B.A. Wooley, A 0.7-V $870 - \mu$W digital-audio CMOS sigma-delta modulator. IEEE J. Solid-State Circuits **44**(4), 1078–1088 (2009)
12. G.C.M. Meijer et al., A three-terminal integrated temperature transducer with micro-computer interfacing. Sens. Actuators **18**, 195–206 (1989)
13. M.A.P. Pertijs, A. Bakker, J.H. Huijsing, A high-accuracy temperature sensor with second-order curvature correction and digital bus interface, in *Proceedings of the ISCAS*, May 2001, pp. 368–371

Chapter 4
BJT-Based, Energy-Efficient Temperature Sensors

As was shown in the previous chapter, the zoom-ADC is well suited for use in energy-efficient temperature sensors. It combines the strengths of SAR- and $\Delta\Sigma$-ADCs to realize an accurate and energy-efficient temperature to digital conversion. In this chapter, a sensor prototype that employs a 1st-order zoom-ADC is described. It is compact and power efficient, requiring only a few μW to operate. Its energy efficiency, however, is limited, due to the use of an inherently slow 1st-order $\Delta\Sigma$ modulator. To improve energy efficiency, a second prototype is presented, which requires less power to operate, while employing a 2nd-order zoom-ADC and a faster sampling scheme. Finally, a third prototype for sensing very high temperatures ($>150\,^\circ$C) is presented, which uses robust techniques to overcome the different sources of temperature sensing errors at such temperatures.

4.1 A Micropower Temperature Sensor [1]

A zoom-ADC can be used to digitize the ratio $X = V_{BE}/\Delta V_{BE}$, which is a measure of temperature. As shown in Fig. 4.1, for $p = 5$, the ratio X is a nonlinear function of temperature, which ranges between 7 and 24 from -40 to $125\,^\circ$C. Once X is known, the PTAT function μ (see Fig. 4.1) can be easily determined as follows:

$$\mu = \frac{\alpha \cdot \Delta V_{BE}}{V_{BE} + \alpha \cdot \Delta V_{BE}} = \frac{\alpha}{X + \alpha}. \tag{4.1}$$

The block diagram of the proposed temperature sensor is shown in Fig. 4.2. It consists of a 1st-order zoom-ADC and an analog front-end: a precision bias circuit whose PTAT output current biases the substrate PNP transistors of a bipolar core. The V_{BE} and ΔV_{BE} voltages extracted from the bipolar core are digitized by the

© Springer International Publishing AG 2018
K. Souri, K.A.A. Makinwa, *Energy-Efficient Smart Temperature Sensors in CMOS Technology*, Analog Circuits and Signal Processing,
DOI 10.1007/978-3-319-62307-8_4

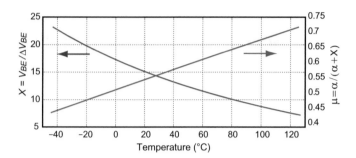

Fig. 4.1 Nonlinear $X = V_{BE}/\Delta V_{BE}(p = 5)$ and linearized $\mu = \alpha/(\alpha + X)$ as a function of temperature

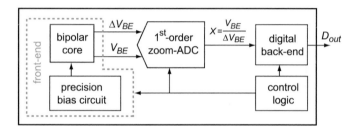

Fig. 4.2 Block diagram of the smart temperature sensor

zoom-ADC, which outputs X to a digital back-end, which, in turn, determines the PTAT function μ and D_{out}, the temperature in degrees Celsius, as given by Eq. (2.9).

4.1.1 Analog Front-End

4.1.1.1 Topology

Figure 4.3 shows the circuit diagram of the analog front-end. Although the temperature dependence of the bias current I_b doesn't impact the accuracy of ΔV_{BE} [see Eq. (2.4)], it does impact the systematic nonlinearity or curvature of V_{BE}, and hence the sensor's systematic error. The curvature in V_{BE} can be reduced by using a PTAT bias current [2]. As shown in Fig. 4.3, the bias circuit uses two PNP transistors, biased at a 5:1 current ratio. The opamp in the feedback loop then forces ΔV_{BE} across the resistor R_b, resulting in an accurate PTAT current I_b, which is then used to bias the PNPs of the bipolar core. Each of the PNPs in the front-end has an emitter area of $A_E = 5\,\mu m \times 5\,\mu m$.

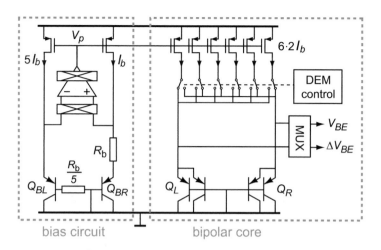

Fig. 4.3 Circuit diagram of the analog front-end

4.1.1.2   Effect of Forward Current Gain β_F

Since a substrate PNP transistor must be biased via its emitter, its collector current and, thus, the resulting V_{BE} will depend on the transistor's current gain β_F. The spread (up to 50%) and temperature dependence of β_F will then impact the accuracy of V_{BE}. This effect becomes more significant as the current gain β_F decreases, as is the case in modern CMOS processes (in the 0.16 μm CMOS process used $\beta_F \approx 4.5$). The technique known as β_F-compensation mitigates this problem by modifying the PTAT bias circuit to generate a β_F-dependent current [3]. This is done by adding a resistor of $R_b/5$ in series with the base of Q_{BL}. The opamp in the feedback loop then ensures that:

$$I_b = \ln(5) \cdot \left(\frac{kT}{q}\right) \cdot \left(\frac{1}{R_b}\right) \cdot \left(\frac{1 + \beta_F}{\beta_F}\right). \tag{4.2}$$

When biased with this emitter current, the collector current of the PNPs in the bipolar core (Q_L, Q_R) will be equal to $\ln(5) \cdot (kT/q)/R_b$, and hence their base-emitter voltages will be insensitive to variations in β_F. However, the accuracy of this technique is limited by current-mirror and β_F mismatch. Therefore, careful layout of the PNPs and the current sources is essential. A further source of V_{BE} error is the current dependency of β_F. In this design, the value of I_b (= 90 nA at 25 °C) was optimized to ensure that both I_b and $5I_b$ are in a relatively flat part of the PNP's β_F versus collector current characteristic (see Fig. 4.4) [2].

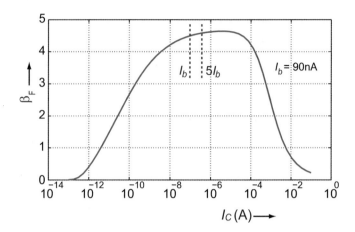

Fig. 4.4 The forward current gain β_F (at $25\,^{\circ}\mathrm{C}$) of a substrate PNP ($A_E = 5\,\mu\mathrm{m} \times 5\,\mu\mathrm{m}$) as a function of collector current I_C in the $0.16\,\mu\mathrm{m}$ CMOS process used

4.1.1.3   Offset Cancellation

Besides the spread in R_b, the offset V_{OS} of the opamp in the bias circuit is a major source of bias current inaccuracy. For the temperature sensor to achieve an inaccuracy of less than $\pm0.2\,^{\circ}\mathrm{C}$, this offset needs to be less than $100\,\mu\mathrm{V}$ [3]. However, in CMOS, this cannot be practically achieved by transistor sizing and careful layout. Therefore, the opamp is chopped, so that the resulting bias current will be switched between $I_b + I_{\text{off}}$ and $I_b - I_{\text{off}}$ where $I_{\text{off}} = V_{OS}/R_b$. Thus, the average of the resulting V_{BE} in Q_L, Q_R will be, to first-order, independent of V_{OS}. However, the use of chopping means that there will be square-wave ripple at the opamp's output V_P. While its amplitude is not important, the complete settling of V_P at the end of each chopping phase is critical, since this is the moment when the zoom-ADC samples the resulting V_{BE}. Due to the large input capacitance of the current-mirror MOSFETs, a typical single-stage high-output-impedance opamp will require a relatively large bias current. In [4] for example, the opamp was chopped at the ADC's sampling rate and drew a large portion of the sensor's supply current ($1.7\,\mu\mathrm{A}$ out of a total of $6\,\mu\mathrm{A}$, or 28%).

4.1.1.4   Opamp Topology

As shown in Fig. 4.5, an adaptive self-biasing opamp [5–7] is used in this work. It consists of a PMOS input pair ($M_{5,6}$) with diode-connected NMOS loads ($M_{1,2}$). Since the opamp's input voltage is chopped, switch S_1 is used to maintain the correct feedback polarity. The voltage output of the input stage ($V_{gs,M1}$ or $V_{gs,M2}$) is converted into a current via M_3 and fed back to the differential pair through the $M_{10}{:}M_{11}$ current mirror. As a result, the tail current of the input stage is derived

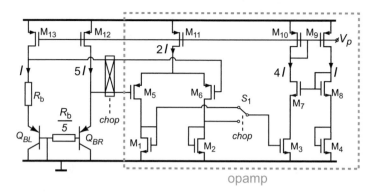

Fig. 4.5 Simplified circuit diagram of the bias current circuit (*left-hand side*) and the positive feedback opamp

from its output voltage. The aspect ratio of the transistors has been chosen such that: $(W/L)_3 = 4 \cdot (W/L)_{1,2}$ and $(W/L)_{10} = 2 \cdot (W/L)_{11}$. This ensures that the current gain of the loop formed by $M_{1,2}{:}M_3$, $M_{10}{:}M_{11}$ is equal to 1 for zero input. When operated in an open-loop configuration, the positive feedback in this current loop would result in ever-increasing/decreasing output currents for negative/positive differential input voltages, corresponding to a very high DC gain. In the bias circuit, however, the amplifier is operated in a negative feedback loop which stabilizes the circuit and enforces a PTAT current I_b. Furthermore, since M_{10} is diode-connected, V_P is a low impedance node, which reduces the time constant associated with the settling of V_P. The opamp's bias current, therefore, can be reduced to meet the relaxed load requirements, while maintaining the gain required.

As any current mirror mismatch will result in input-referred offset, high overdrive voltages (260 mV and 130 mV for the PMOS and NMOS devices respectively, at 25 °C) and careful layout are essential. To minimize the effect of channel length modulation on the current gain of the positive feedback loop, a replica circuit drives M_8 and ensures that the V_{ds} of M_3 is equal to that of $M_{1,2}$. At 25 °C, the opamp draws only 630 nA, significantly less (63%) than in our previous work [4]. The entire front-end draws only 2.1 μA from a 1.8 V supply.

4.1.1.5   Precision Issues

Mismatch in current sources and bipolar devices impacts the accuracy of ΔV_{BE}. Even with careful layout, a relative current ratio mismatch $\Delta p/p$ in the order of 0.1% can be expected, at best. For $p = 5$, this will lead to an error of about 0.14 °C at $T = 25$ °C [2]. Therefore, dynamic element matching (DEM) of the six current sources and two bipolar transistors in the bipolar core is essential to generate the accurate 1:5 current ratio required for an accurate ΔV_{BE}. As shown in Fig. 4.3, DEM switches are placed in series with each current source, in order to swap their current direction according to the DEM control logic. Since V_{BE} and ΔV_{BE} are sampled at the emitter terminal of the BJTs, the voltage drop across DEM switches doesn't impact the accuracy of the sampled voltages.

4.1.2 Zoom ADC

4.1.2.1 Topology

In this work, a 1st-order zoom-ADC is used to digitize the ratio $X = V_{BE}/\Delta V_{BE}$. As previously discussed, it combines a coarse SAR-ADC and a fine 1st-order $\Delta\Sigma$ converter in a two-step conversion scheme [4, 5]. In this topology, the digital ratio X is accurately resolved, with a high resolution and within a fairly short conversion time.

As shown in Fig. 4.1, the ratio $X = V_{BE}/\Delta V_{BE}$ ranges from 7 to 24 from -40 to 125 °C ($p = 5$). X can thus be expressed as $X = n + \mu'$, where n and μ' are its integer and fractional parts, respectively, and can be determined separately, as shown in Fig. 4.6. In a coarse conversion, n is determined by a SAR algorithm, which compares V_{BE} to integer multiples of ΔV_{BE}. In a succeeding fine conversion, the fraction μ' is then determined by a 1st-order $\Delta\Sigma$-ADC, whose references are chosen so as to zoom into the region determined by the SAR algorithm, i.e., from $n \cdot \Delta V_{BE}$ to $(n + 1) \cdot \Delta V_{BE}$. The region of interest is now quite small (less than 18 °C), thus relaxing the resolution requirement on the $\Delta\Sigma$-ADC, which in turn leads to energy-efficient conversions [4, 5].

4.1.2.2 Implementation

As shown in Fig. 4.7, the zoom-ADC is basically a modified 1st-order switched-capacitor (SC) $\Delta\Sigma$-ADC with 24 sampling capacitors. At the start of each comparison step of the coarse conversion, the integrator is reset, and it therefore functions as a sample-and-hold circuit. As shown in Fig. 4.8b, V_{BE} is then sampled on a

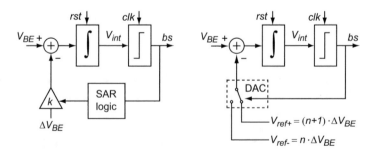

Fig. 4.6 Block diagram of the 1st-order zoom-ADC during the coarse (*left*) and fine (*right*) conversions

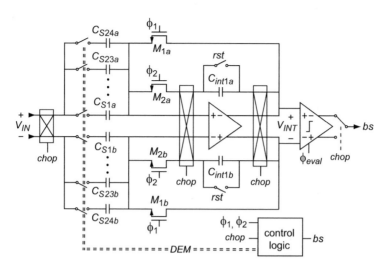

Fig. 4.7 Simplified circuit diagram of the 1st-order zoom-ADC

single unit capacitor and integrated during one full clock cycle. In the next clock cycle, $-\Delta V_{\mathrm{BE}}$ is sampled on k unit capacitors and also integrated, thus a total charge proportional to $(V_{\mathrm{BE}} - k \cdot \Delta V_{\mathrm{BE}})$ is integrated. The comparator's output bs then indicates the result of the comparison $(V_{\mathrm{BE}} > k \cdot \Delta V_{\mathrm{BE}})$. The control logic implements the SAR algorithm, with which n can be determined within five comparison steps, since $n \leq 24$. Once n is known, the fine conversion step is determined with a $\Delta\Sigma$ charge-balancing scheme (Fig. 4.6). After an initial reset, the modulator operates as follows: when bs $= 0$, $(V_{\mathrm{BE}} - n \cdot \Delta V_{\mathrm{BE}})$ is integrated, and when bs $= 1$, $(V_{\mathrm{BE}} - (n + 1) \cdot \Delta V_{\mathrm{BE}})$ is integrated. As shown in Fig. 4.8, such integrations require two clock cycles: one to integrate V_{BE} and one to integrate $-k \cdot \Delta V_{\mathrm{BE}}$. Since the net integrated charge is approximately zero, the bitstream average is the desired $\mu = (V_{\mathrm{BE}} - n \cdot \Delta V_{\mathrm{BE}})/\Delta V_{\mathrm{BE}}$.

When $V_{\mathrm{BE}} \sim k \cdot \Delta V_{\mathrm{BE}}$, non-idealities such as comparator offset, noise, and mismatch during the coarse conversion could lead to incorrect n values, and therefore clipping in the fine conversion. As discussed in Chap. 3, this can be avoided by performing an extra guard-band cycle. In this work, the fine conversion is appropriately extended to $3 \cdot \Delta V_{\mathrm{BE}}$, thus relaxing the requirements on the coarse conversion [5]. As shown in Fig. 4.9, the range is set-up to cover from $(n-1) \cdot \Delta V_{\mathrm{BE}}$ to $(n + 2) \cdot \Delta V_{\mathrm{BE}}$, in such a way that V_{BE} is always roughly in the middle of this range, and hence avoiding any out-of-ranging in the fine conversion step. In order to accommodate the required accuracy of the gain factor k, DEM is applied to the elements of the sampling cap-DAC during the $\Delta\Sigma$ conversion step.

As illustrated in Fig. 4.7, the main element of the zoom-ADC is a SC integrator built around a fully differential folded-cascoded opamp with a gain of 86 dB, which is required to maintain the linearity of overall ADC across the fine conversion segments. However, due to the relaxed requirements on the ADC's resolution, no

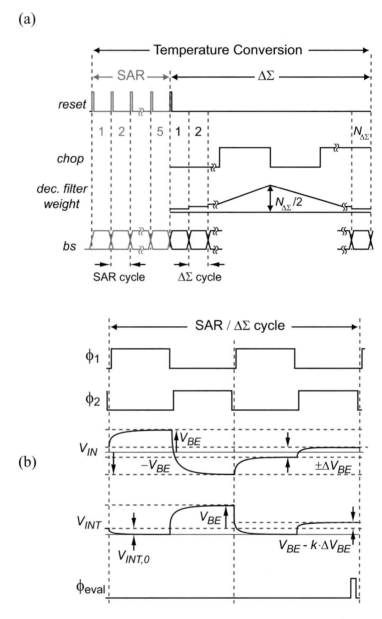

Fig. 4.8 (a) Timing diagram of a temperature conversion: (b) Waveforms of a full SAR $\Delta\Sigma$ cycle. V_{IN} zoom ADC's input voltage, V_{INT} integrator's output voltage, $V_{INT,0}$ integrator's initial voltage. The gain factor k is set by the SAR-logic in the coarse conversion, while $k = n$ or $n + 1$ when bs $= 0$ or 1 in the fine conversion step

Fig. 4.9 The range is
extended to $3 \cdot \Delta V_{BE}$ to avoid
out-of-range errors during the
fine conversion step

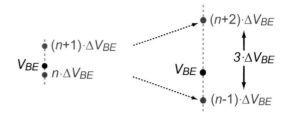

gain boosting is required, unlike [3, 8], thus reducing the area and power of the opamp. A SC common-mode feedback scheme is applied to maintain the opamp's output common mode at a well-defined value ($\approx V_{DD}/2$). The sampling capacitors are also quite small: $C_S = 120$ fF, while the integration capacitors are $2 \cdot C_S$, in order to maintain the integrator swing within the folded-cascode topology limits. The size of sampling caps are defined based on the target resolution at sensor's output, which is determined by kT/C noise of the sampling capacitors, and the conversion time.

The opamp's offset and $1/f$ noise are reduced by correlated double-sampling (CDS) during both the coarse and fine conversions. Another source of inaccuracy in a SC integrator is the charge injection of the switches. A fully differential structure combined with the bottom-plate sampling technique is known to be the most effective way to mitigate this, as the injected charge then appears as a common-mode effect [9]. The effectiveness of this technique, nevertheless, is mainly limited by the mismatch between switches, and thus their charge injection. To overcome this, the entire ADC is chopped twice per fine conversion. This is done by swapping the polarity of input voltage and the quantizer output, i.e., bitstream bs. State preserving switches around the opamp (Fig. 4.7) are essential to maintain the integrator state when chopping is applied [3]. The timing diagram of the chopping signal is shown in Fig. 4.8a.

4.1.3   Measurement Results

The temperature sensor was realized in a standard $0.16\,\mu$m CMOS process with five metal layers (Fig. 4.10). The chip has an active area of $0.12\,\text{mm}^2$, and consumes $8.2\,\mu$W from a 1.8 V supply at 25 °C. The digital back-end, the control logic, and the fine conversion's sinc^2 decimation filter were implemented off-chip for flexibility. For characterization, the prototypes were packaged in ceramic DIL packages and placed in a climate chamber, in good thermal contact with an aluminum block containing a platinum Pt-100 resistor calibrated to 20 mK. With this setup, 19 samples from one batch were then characterized over the temperature range from -30 to 125 °C. As shown in Fig. 4.11, the resulting batch-calibrated inaccuracy was ± 0.5 °C (3σ), after digital compensation for residual curvature (± 0.25 °C). The standard deviation σ is an estimated value, which is obtained based on a limited number of samples: 19 samples in this case. A single digital trim at 25 °C was used

Fig. 4.10 Chip micrograph of the first prototype sensor employing a 1st-order zoom-ADC

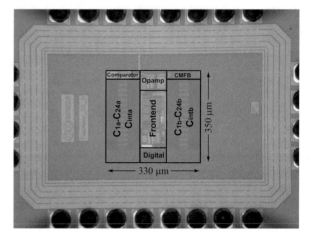

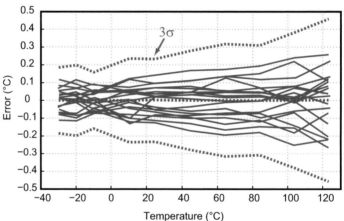

Fig. 4.11 Measured temperature error of 19 sensors before trimming; *dashed lines* refer to the average and $\pm 3\sigma$ limits

to compensate for V_{BE}'s PTAT spread [4, 5]. This was done by individually tuning and embedding the α value for each sensor in the digital back-end, thereby reducing the inaccuracy to $\pm 0.2\,^{\circ}$C (3σ) as shown in Fig. 4.12. At 10 conversions/s (1024 $\Delta\Sigma$ cycles), the sensor achieves a resolution of 15 mK (rms). The sensor operates from a 1.6 to 2 V supply with a sensitivity of 0.1 $^{\circ}$C/V. The sensor's performance is summarized in Table 4.1 and compared to other accurate temperature sensors [3, 8].

To assess the sensor's energy efficiency, its resolution figure-of-merit (FoM) can be calculated. The sensor operates from a minimum supply voltage of 1.6 V and draws 4.6 μA. Given the resolution of 15 mK (rms) in a conversion time of 100 ms, this translates into a resolution FoM of 0.17 nJ $^{\circ}$C^2. Compared to the other designs in Table 4.1, the energy efficiency has been improved by more than 11×, while achieving comparable accuracy.

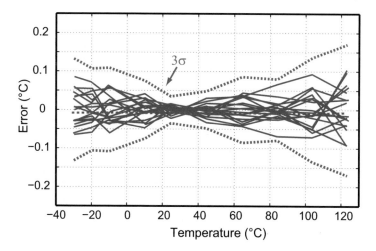

Fig. 4.12 Measured temperature error of 19 sensors after a single α-trim at 25 °C; *dashed lines* refer to the average and $\pm 3\sigma$ limits

Table 4.1 Performance summary of the first prototype employing a 1st-order zoom-ADC and compared to previous work

Parameter	First prototype JSSC'11 [1]	JSSC'05 [3]	ISSCC'09 [8]
CMOS technology	0.16 μm	0.7 μm	0.7 μm
Chip area	0.12 mm^2	4.5 mm^2	4.5 mm^2
Supply current (RT)[a]	4.6 μA	75 μA	25 μA
Supply voltage	1.6–2 V	2.5–5.5 V	2.5–5.5 V
Inaccuracy (trim points)	± 0.2 °C (3σ) (1)	± 0.1 °C (3σ) (1)	± 0.1 °C (3σ) (1)
Temperature range	-30 to 125 °C	-55 to 125 °C	-55 to 130 °C
Resolution (T_{conv})	0.015 °C (100 ms)	0.01 °C (100 ms)	0.025 °C (100 ms)
Resolution FoM	0.17 nJ °C^2	1.9 nJ °C^2	3.9 nJ °C^2

[a]Excluding the off-chip digital

4.2   An Energy-Efficient Temperature Sensor [10]

As demonstrated in the previous section, using a 1st-order zoom-ADC results in a low-power sensor. However, its energy efficiency is limited by the low conversion rate of its 1st-order modulator. Furthermore, to maintain linearity across the zoom segments, an opamp with a DC gain in excess of 80 dB was necessary. Combined with the large swing requirement ($>\pm 350$ mV) in the V_{BE} integration phase (see Fig. 4.8), this led to a topology with limited power efficiency in the target 0.16 μm CMOS process. In this section a temperature sensor based on a 2nd-order zoom-ADC will be presented that is significantly more energy-efficient, and achieves the required loop-gain while dissipating less power.

4.2.1 Improving Energy Efficiency

As in the first prototype, a 5-bit SAR-ADC performs the coarse conversion step. The fine step, however, uses the block diagram of Fig. 4.13. It is based on a single-bit feed-forward 2nd-order $\Delta\Sigma$-ADC. This architecture requires about 8× less $\Delta\Sigma$ cycles to achieve almost the same resolution as in the first prototype. Moreover, the overall loop-gain is now achieved by using two low-gain, low-power integrators, thus significantly improving power and energy efficiency.

The modulator's stability is achieved by a single feed-forward path around the second integrator. This leverages a distinct advantage of the zooming algorithm: the fact that since $V_{BE} \sim n \cdot \Delta V_{BE}$ during the fine conversion step, the error signal processed by the loop filter is quite small, thus reducing the output swing of the two opamps. As a result, no extra direct feed-forward path between the input terminal and the quantizer's input is required, as is the case in the well-known low-swing feed-forward architecture [11]. This simplifies the modulator's implementation, reduces the loading on the analog front-end that generates V_{BE}, and eliminates a potential source of parasitic coupling into the summing node at the quantizer's input.

As previously discussed, the range in the fine conversion step of the first prototype was extended by 3× to accommodates small errors during the coarse conversion phase. In this work, the range is extended by only 2×, which further improves the conversion speed, since the modulator's full range has been reduced. The necessary information is obtained during a guard-band step, in which V_{BE} is compared to $(n + 0.5) \cdot \Delta V_{BE}$. Depending on the result, the references of the $\Delta\Sigma$-ADC are then set to either $(n - 1) \cdot \Delta V_{BE}$ and $(n + 1) \cdot \Delta V_{BE}$, or $n \cdot \Delta V_{BE}$ and $(n + 2) \cdot \Delta V_{BE}$. In the rest of this section, for simplicity, we shall assume that the former is the case, as shown in Fig. 4.13.

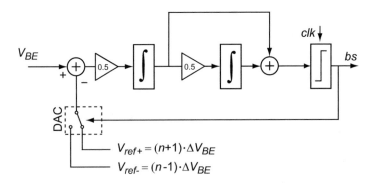

Fig. 4.13 Block diagram of the proposed 2nd-order zoom-ADC during the fine conversion

4.2.2 An Energy-Efficient Integration Scheme

During the fine conversion, as shown in Fig. 4.13, every $\Delta\Sigma$ cycle requires the integration of either $(V_{BE} - (n-1) \cdot \Delta V_{BE})$ or $(V_{BE} - (n+1) \cdot \Delta V_{BE})$, when the comparator's output bs is either 0 or 1, respectively. For simplicity, let's assume that a charge proportional to $(V_{BE} - k \cdot \Delta V_{BE})$ is integrated during one $\Delta\Sigma$ cycle, where k is either $(n-1)$ or $(n+1)$ depending on the polarity of bs. In the first prototype, this was performed using a SC integrator and in two clock cycles: in a first clock cycle a charge proportional to V_{BE} was integrated, while in a second clock cycle a charge proportional to $-k \cdot \Delta V_{BE}$ was integrated. A folded-cascode opamp was therefore required to accommodate the large swing during the V_{BE} integration phase, and thus the low-swing advantage of zooming was not fully exploited.

In this design, the two clock cycles are combined, i.e., both V_{BE} and ΔV_{BE} are simultaneously sampled and then integrated in only *one* clock cycle. As shown in Fig. 4.14 during the sampling phase ϕ_1, V_{BE} is sampled on C_S while $-\Delta V_{BE}$ is simultaneously sampled on $k \cdot C_S$, thus a charge proportional to $(V_{BE} - k \cdot \Delta V_{BE})$ is stored on the sampling capacitors. The polarity of both input voltages is swapped during ϕ_2, and therefore a charge proportional to $2 \cdot (V_{BE} - k \cdot \Delta V_{BE})$ is integrated during each clock cycle. Due to the charge cancellation between V_{BE} and $-k \cdot \Delta V_{BE}$,

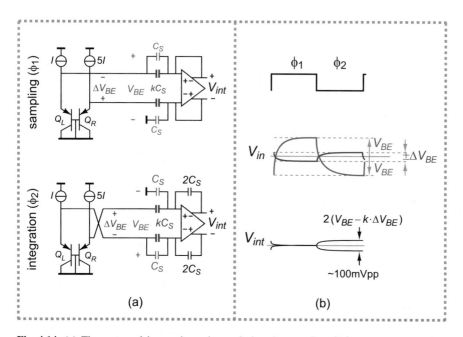

(a) (b)

Fig. 4.14 (a) The proposed integration scheme during the sampling (ϕ_1) and integration (ϕ_2) phases. (b) Waveforms of a full SAR/$\Delta\Sigma$ cycle. V_{IN} zoom-ADC's input voltage, V_{INT} integrator's output voltage. The gain factor k is set by the SAR-logic in the coarse conversion, while $k = n-1$ or $n+1$ when bs = 0 or 1 in the fine conversion step

the integrated charge difference is quite small, and can be accommodated by a low-swing, and power-efficient telescopic opamp. Moreover, this approach also halves the conversion time, thus improving the energy efficiency by another factor of 2.

4.2.3   Implementation

4.2.3.1   Circuit Diagrams

Figure 4.15 shows the simplified circuit level diagram of the proposed 2nd-order zoom-ADC. A capacitor DAC with 28, 120 fF unit capacitances realizes the gain factor k required for ΔV_{BE} sampling, while an extra capacitor $C_G = 0.5 \cdot C_S$ is used during the guard-band step. To simultaneously sample V_{BE} and in order to cover the military temperature range, the number of unit elements in the capacitor DAC is increased to 29. During the coarse conversion step, a switch S_{bp} bypasses the second integrator, thus directly connecting the output of the first integrator to the comparator. Moreover, at the start of each comparison step, the first integrator is reset and therefore acts as a sample-and-hold.

The first integrator is built around a power-efficient, fully differential telescopic opamp, which only draws 600 nA, has a gain of 76 dB, and a maximum swing of about ± 200 mV. As shown in Fig. 4.15, a pseudo-differential inverter-based OTA forms the second integrator [12]. At 25 °C, it draws 140 nA, occupies only

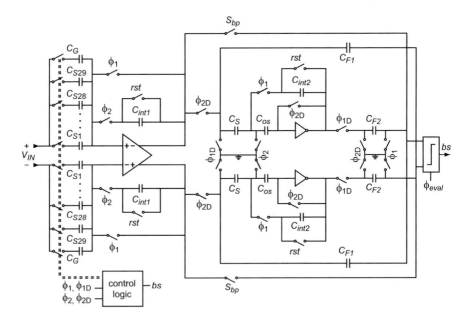

Fig. 4.15 Simplified circuit diagram of the proposed 2nd-order zoom-ADC

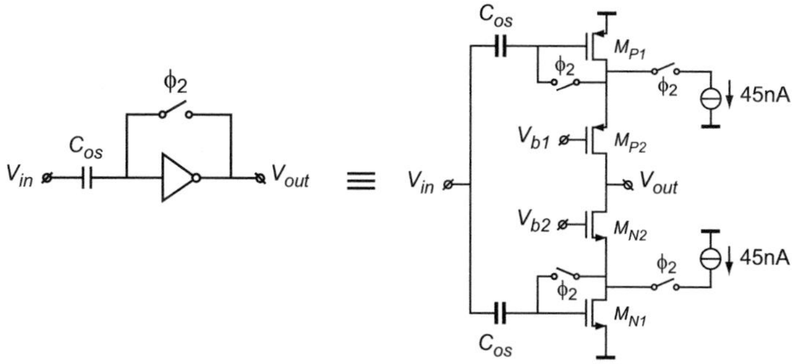

Fig. 4.16 Circuit diagram of the proposed inverter-based OTA

$0.002\,\text{mm}^2$, and has a gain of $\sim$44 dB. During ϕ_2, when the output of first integrator is sampled on capacitors C_S, the two inverter-based OTAs are in unity-gain configuration and auto-zeroed via offset storing capacitors C_{OS}. Due to the feedback path through the integration capacitors in ϕ_1, a virtual ground is formed, thus pushing the sampled charge into the integration capacitors $C_{int2} = 2 \cdot C_S$. Figure 4.16 shows the implementation details of the inverter-based OTA. To decrease the inverter's sensitivity to power supply and process spread, a dynamic current-biasing technique is proposed. During the auto-zeroing phase ϕ_2, M_{N1} and M_{P1} are diode-connected and biased with two current sources (45 nA each), while their operating bias voltages are stored on offset storing capacitors C_{OS}. The bias voltages V_{b1} and V_{b2} are chosen such that M_{N2}, M_{P2} are essentially off during ϕ_2. The two currents are mirrored from the front-end's precision bias circuit to ensure robustness to supply and process variations. After disconnecting the two current sources in ϕ_1, M_{N1} and M_{P1} are configured as common-source and form a class-AB amplifier, with a virtual ground at V_{IN}. Since the output voltage swing requirement is reduced to about $\pm 100\,\text{mV}$ by the prior coarse conversion step, cascoding of M_{N1} and M_{P1} is readily possible, thus enhancing the inverter's output resistance, and hence its DC gain. A passive summation network at the input of quantizer combines the output of second integrator with that of the first integrator via the feed-forward capacitor C_{F1}, as shown in Fig. 4.15. To set-up the various biasing voltages, the ADC requires a start-up time of $120\,\mu\text{s}$ (three clock cycles) before each conversion.

4.2.3.2   Precision Techniques

During the fine conversion, the accuracy of the ratio k is determined by the matching between the unit capacitor that samples V_{BE} and the k capacitors which sample ΔV_{BE}. Any mismatch will lead to a nonlinear ADC transfer function. The matching of the references should, therefore, be commensurate with the ADC's target resolution, i.e., 13 bits. Since this cannot be achieved by layout alone, a dynamic element matching (DEM) scheme was used.

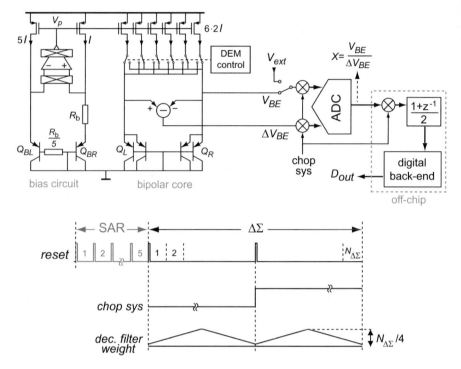

Fig. 4.17 *Top*: Block diagram of the temperature sensor. *Bottom*: Timing diagram of a full temperature conversion

Figure 4.17 shows the block diagram of the sensor and the timing of a full temperature conversion. The analog front-end consists of a bias circuit and a bipolar core. As in the first prototype, the bias circuit generates a PTAT current $I = 90$ nA (at 25 °C) with the help of a low power, self-biased chopped opamp and two auxiliary PNPs [1]. As in the first prototype, a β_F-compensation technique is employed to ensure that Q_R and Q_L are biased with β_F-independent collector currents, thus improving the robustness of the resulting V_{BE} to process spread. Furthermore, the six current sources and the two bipolar transistors in the bipolar core are dynamically matched to achieve (on average) the accurate 1:5 current ratio required to generate an accurate ΔV_{BE}. As in [1], each of the PNPs in the front-end has an emitter area of $A_E = 5\,\mu m \times 5\,\mu m$.

As previously discussed in Sect. 4.1.2.2, a major source of inaccuracy, the offset, and $1/f$ noise of integrators can be reduced by employing correlated-double sampling (CDS) during the coarse and fine conversions. In order to minimize the effect of charge injection, both integrators use differential topologies with minimum-size switches around the integration capacitors. In contrast to the first prototype, however, a digital rather than an analog implementation of system-level chopping is employed. As shown in Fig. 4.17, after an initial coarse conversion, the $\Delta\Sigma$ conversion is performed twice with swapped input voltage polarities, and the two digital results are then averaged [12]. This eliminates the need for state-preserving choppers around the integration capacitors, which simplifies the layout and eliminates a potential source of charge injection, which could otherwise cause ADC nonlinearity. Compared to the conventional analog approach, however, this results in a small loss in resolution: up to 0.5 bits if the ADC is quantization-noise limited.

4.2.4 Realization and Measurements

The sensor was realized in a standard $0.16\,\mu m$ CMOS process with five metal layers and has an active area of $0.08\,mm^2$, as shown in Fig. 4.18. For flexibility, the digital back-end, the control logic, and the fine conversion's $sinc^2$ decimation filter [13] were implemented off-chip. At $25\,°C$, the sensor draws $3.4\,\mu A$ and operates from a 1.5 to 2 V supply with a supply sensitivity of $0.5\,°C/V$. Running at a clock frequency of $25\,kHz$, it requires a conversion time of $5.3\,ms$ ($128\ \Delta\Sigma$ cycles) to achieve a kT/C limited resolution of $20\,mK$ (rms), which improves to about $5\,mK$ (rms) if the conversion time is extended to $100\,ms$. For characterization, 18 devices from one batch were packaged in ceramic DIL packages and measured over the military temperature range from -55 to $125\,°C$. As shown in Fig. 4.19, the resulting inaccuracy after batch calibration was $\pm0.6\,°C$ (3σ), with a residual curvature of only $\pm0.03\,°C$, which is significantly less than that of the first prototype ($\pm0.25\,°C$). This was achieved by optimizing the design and layout of the Cap-DAC, in order to minimize the cross-coupling between different unit elements, thereby enhancing the overall linearity of the converter. The Cap-DAC layout is also very compact, as can be seen by comparing Fig. 4.18 to Fig. 4.10. To further improve the sensor's accuracy, individual calibration and trimming are essential. In the following, two different approaches based on thermal and electrical measurements are presented.

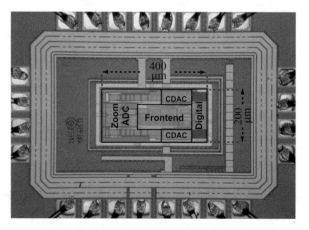

Fig. 4.18 Chip micrograph of the second prototype sensor employing a 2nd-order zoom-ADC

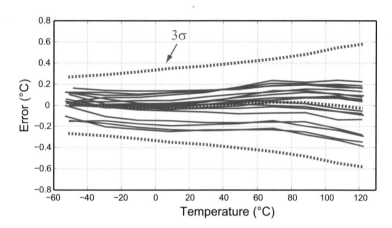

Fig. 4.19 Measured temperature error of 18 sensors before trimming; *dashed lines* refer to the average and $\pm 3\sigma$ limits

4.2.5 Thermal Calibration

Individual calibration of an integrated temperature sensor requires accurate information about its die temperature. Conventionally, this is obtained by bringing the device under test (DUT) and a reference temperature sensor to exactly the same temperature, whereupon the outputs of both devices are logged. As for the first prototype, the reference sensor is a platinum Pt-100 resistor calibrated to an inaccuracy of 20 mK. Both sensors are embedded in a large metal block, which acts as a thermal low-pass filter and facilitates measurements with milli-Kelvin stability [14].

Three different single-parameter trimming methods were investigated. First, for each sensor, the offset parameter B in Eq. (2.9) was adjusted so as to cancel the error at the calibration temperature (30 °C). After this offset trim, the sensor's inaccuracy is less than $\pm 0.25\,°C$ (3σ) from -55 to 125 °C. Alternatively, the parameter α in Eq. (3.1) can be adjusted, as was done in the first prototype [1]. The resulting inaccuracy, however, is almost exactly the same as that obtained with offset trim. Since the dominant source of sensor inaccuracy, i.e., the spread in V_{BE} is PTAT in nature (see Fig. 4.19), a digital PTAT trim [2], as discussed in Sect. 3.4 is also employed. The resulting inaccuracy is then less than $\pm 0.15\,°C$ (3σ), as shown in Fig. 4.20.

4.2.6 Voltage Calibration

Although thermal calibration can be performed very accurately, the long stabilization time required for the DUT and the reference sensor to reach thermal equilibrium prohibits its use as a low-cost calibration method. In [15], a voltage calibration

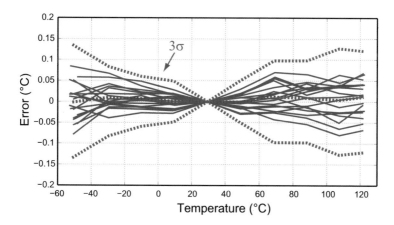

Fig. 4.20 Measured temperature error of 18 sensors after thermal calibration and PTAT trimming at 30 °C; *dashed lines* refer to the average and $\pm 3\sigma$ limits

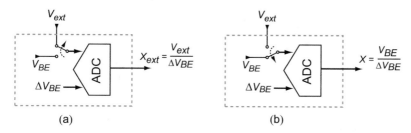

Fig. 4.21 Voltage calibration requires two ADC conversions: one to obtain the actual die temperature (**a**) and another to obtain the untrimmed output (**b**)

method was proposed, in which die temperature is established by measuring an on-chip ΔV_{BE}. By applying DEM to the six current sources and the two PNPs (see Fig. 4.17), the collector current ratio p, and therefore ΔV_{BE} can be made robust to process spread. The process-dependent non-ideality factor η (= 1.0042) can also be extracted by batch calibration. As shown in Fig. 4.21, the die temperature can then be determined by the following procedure. First, V_{BE} is replaced by an accurate external voltage V_{ext} (see Fig. 4.21a). The on-chip ADC then digitizes the ratio $X_{ext} = V_{ext}/\Delta V_{BE}$ accurately and with high resolution, whereupon the actual die temperature T_D can be calculated:

$$\Delta V_{BE} = \eta \cdot \frac{kT_D}{q} \cdot \ln(p), X_{ext} = \frac{V_{ext}}{\Delta V_{BE}} \Rightarrow T_D = \frac{V_{ext}}{C_m \cdot X_{ext}}, C_m = \eta \cdot \frac{k}{q} \cdot \ln(p).$$
(4.3)

In a second step, V_{ext} is replaced by the on-chip V_{BE} and a normal conversion is performed to determine $X = V_{BE}/\Delta V_{BE}$, and hence the sensor's untrimmed output (see Fig. 4.21b). In contrast to thermal calibration, this approach can be performed at room temperature, and is much faster, requiring only two ADC conversions. Since

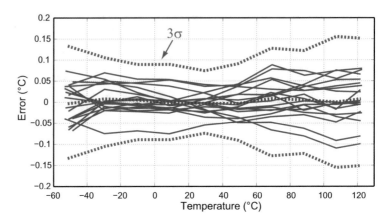

Fig. 4.22 Measured temperature error of 18 sensors after voltage calibration and PTAT trimming at room temperature; *dashed lines* refer to the average and $\pm 3\sigma$ limits

the sensor achieves a resolution of 5 mK in a conversion time of 100 ms, which is commensurate with the expected $\pm 0.15\,^\circ$C inaccuracy, this means that the total calibration time is only 200 ms.

Compared to the results of thermal calibration, the results of voltage calibration followed by an offset or digital PTAT trim are only slightly worse around room temperature. The worst-case inaccuracy from -55 to $125\,^\circ$C, however, is almost exactly the same as shown in Fig. 4.22. This confirms the fact that the inaccuracy of ΔV_{BE} is negligible, and so voltage calibration is a robust alternative to thermal calibration.

4.2.7   Batch-to-Batch Spread and Plastic Packaging

To verify the effect of batch-to-batch spread on sensor inaccuracy, devices from a different process batches were characterized. As before, 18 devices from one batch were packaged in ceramic DIL packages and measured from -55 to $125\,^\circ$C. Table 4.2 compares the resulting inaccuracy and calibration parameters (i.e., A, B, and α) of the two batches. As shown, the resulting inaccuracy after batch calibration has increased to $\pm 0.25\,^\circ$C (3σ). Moreover, the obtained gain and offset parameters A, B after batch calibration show a batch-to-batch spread of about 0.4% and 0.3%, respectively. Since at room temperature the PTAT ratio $\mu \approx 0.5$, this translates to a temperature shift of about $-0.5\,^\circ$C.

However, the optimal mapping coefficient α changes by less than 0.1% from batch-to-batch. According to the discussion in Sect. 3.4, such a small variation results in a minimal impact on the parameters A and B, and so α can be regarded as a digital constant. Finally, the non-ideality factor η only changes by about 0.02% from batch-to-batch, which corresponds to a maximum calibration error of about 50 mK.

Table 4.2 Impact of batch-to-batch spread on sensor accuracy and calibration parameters

	Batch-1	Batch-2
Untrimmed inaccuracy (3σ)	$\pm 0.6\,°C$	$\pm 0.6\,°C$
PTAT-trimmed inaccuracy (3σ)	$\pm 0.15\,°C$	$\pm 0.25\,°C$
α (mapping coefficient)	15.44	15.45
A (gain parameter)	613.31	610.74
B (offset parameter)	283.70	282.93
η (non-ideality factor)	1.0042	1.0044

Table 4.3 Effect of mechanical stress on sensor accuracy and calibration parameters

	Ceramic package	Plastic package
Untrimmed inaccuracy (3σ)	$\pm 0.6\,°C$	$\pm 0.8\,°C$
PTAT-trimmed inaccuracy (3σ)	$\pm 0.25\,°C$	$\pm 0.25\,°C$
α (mapping coefficient)	15.45	15.47
A (gain parameter)	610.74	611.59
B (offset parameter)	282.93	283.94
η (non-ideality factor)	1.0044	1.0044

In production, low-cost plastic packages are preferred to ceramic packages. The associated mechanical stress, however, impacts the sensor's accuracy, an effect which is referred to as packaging shift [16], and results in a fairly systematic modification to the base-emitter voltage V_{BE} [17, 18]. To evaluate this, 22 samples from the same batch of the second batch were packaged in plastic DIP packages and then characterized. As shown in Table 4.3, the untrimmed inaccuracy after batch calibration increased to about $\pm 0.8\,°C$ (3σ). However, a PTAT trim reduced the inaccuracy to about $\pm 0.25\,°C$ (3σ), which is equivalent to that obtained with ceramic packaging. The optimal mapping coefficient α changed by about 0.2%, while the fitting parameters A, B changed by about 0.15% and 0.35%, respectively, which corresponds to a packaging shift of about $-0.36\,°C$ at room temperature.

From these measurements, it can be concluded that in order to achieve high accuracy over different batches and different packages, batch calibration is essential. Once the fitting parameters A and B are known, individual devices can be trimmed on the basis of a fast voltage calibration, since the non-ideality factor appears to be essentially constant over different batches and packages.

4.2.8 Noise and ADC Characteristics

As in other two-step ADC structures, mismatch between the references used in the various fine conversion steps could result in discontinuities in the ADC's characteristic. To examine this, the ADC's input range was swept by slowly sweeping the oven temperature from -40 to $100\,°C$ over a 3 h period, while continuously logging

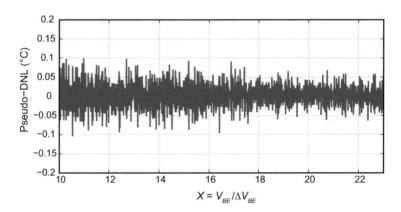

Fig. 4.23 Measured pseudo-DNL versus X. The sensor's conversion time is 5.3 ms

the sensor's output. This corresponds to a temperature slope of ≈ 13 mK/s, which implies that the temperature change between successive measurements is less than 1 mK, i.e., much smaller than the sensor's own resolution. Taking the difference between successive sensor outputs then results in a pseudo-DNL function, which reflects the ADC's resolution and possible discontinuities between the various fine conversion segments. As shown in Fig. 4.23, the sensor achieves a resolution of 20 mK (rms) into 5.3 ms around room temperature ($X \approx 14.5$), which is enough to calibrate it rapidly to $\pm 0.2\,^\circ$C inaccuracy. Moreover, there are no discontinuities between the different fine segments. Lastly, it can be seen that the sensor's resolution is slightly temperature dependent. This is due to the fact that the full-scale range of each fine conversion is not constant, but is equal to $2 \cdot \Delta V_{BE}$.

4.2.9   Comparison to Previous Work

The sensor's performance is summarized in Table 4.4 and compared to the first prototype as well as to other energy-efficient, low power state-of-the-art temperature sensors, with digital output. It employs a low-cost, room-temperature voltage calibration technique, and it also achieves the highest accuracy over a wide temperature range: $\pm 0.15\,^\circ$C (3σ) from -55 to $125\,^\circ$C. Compared to the first prototype, this work achieves comparable resolution in about $18\times$ less conversion time, while consuming 25% less supply current, thus improving the energy efficiency by over $20\times$. This is also in line with the performance of state-of-the-art thermistor- and MOSFET-based sensors, and is evidenced by a resolution FoM of 11 pJ $^\circ$C^2, which is the best among others.

Table 4.4 Performance summary and comparison to previous work

Parameter	Second prototype JSSC'13 [10]	First prototype JSSC'11 [1]	ISSCC'14 [19]	JSSC'05 [20]	ISSCC'09 [21]
Sensor type	PNP	PNP	NPN	Resistor	MOSFET
CMOS technology	0.16 µm	0.16 µm	0.18 µm	0.18 µm	0.18 µm
Chip area	0.08 mm^2	0.12 mm^2	0.085 mm^2	0.18 mm^2	0.032 mm^2
Supply current (RT)[a]	3.4 µA	4.6 µA	4.5 µA	20 µA	0.4 µA
Supply voltage	1.5–2 V	1.6–2 V	1.4–2 V	1.2–2 V	0.9–1.1 V
Supply sensitivity	0.5 °C/V	0.1 °C/V	–	0.625 °C/V	8 °C/V
Inaccuracy	±0.15 °C (3σ)	±0.2 °C (3σ)	–	±0.5 °C (max)	−0.8/ + 1 °C (max)
Temperature range	−55 to 125 °C	−30 to 125 °C	−45 to 85 °C	0 to 100 °C	0 to 100 °C
Calibration (points)	Voltage (1)	Thermal (1)	–	Thermal (1)	Thermal (2)
Resolution (T_{conv})	0.02 °C (5.3 ms)	0.015 °C (100 ms)	0.025 °C (6 ms)	0.25 °C (12.5 µs)	0.3 °C (1 ms)
Resolution FoM	11 pJ °C^2	170 pJ °C^2	24 pJ °C^2	19 pJ °C^2	32 pJ °C^2

[a]Excluding the off-chip digital

4.3   Sensing High Temperatures [22]

While most applications only require temperature sensors that operate up to 125 °C, automotive and industrial applications require operation at much higher temperatures (>150 °C). Achieving this in CMOS is not a trivial task and so off-chip sensors such as thermistors and thermocouples are often used. The main challenge lies in the fact that the leakage and saturation currents of CMOS components increase rapidly at high temperatures, which, in turn, leads to significant temperature-sensing errors. While dissipating 65 µW, the sensor in [23] achieves an inaccuracy of ±0.1 °C (3σ) from −55 to 125 °C after a one-point trim. However, its inaccuracy increases to about ±1 °C at 150 °C. While drawing 180 µA and after a two-point trim, the sensor in [24] achieves an inaccuracy of ±0.5 and ±2.5 °C from −55 to 175 °C in the current and voltage modes of operation, respectively. In a commercially available product, an inaccuracy of ±1 °C was achieved up to 175 °C (after trimming) [25]. In another product, an inaccuracy of ±3 °C was achieved at 200 °C [26].

As explained in Chap. 1, CMOS temperature sensors based on the thermal diffusivity (TD) of silicon generate a temperature-dependent phase-shift, and are thus insensitive to leakage currents. This allows them to operate up to very high temperatures without compromising their accuracy. In [27], an inaccuracy of $\pm 0.4\,^{\circ}\mathrm{C}$ (3σ) from -70 to $200\,^{\circ}\mathrm{C}$ was achieved with a one-point trim. However, the sensor's operation requires the generation of heat pulses, resulting in a power dissipation of 2.6 mW.

In this section, a modified version of the second prototype is presented, which achieves an inaccuracy of $\pm 0.4\,^{\circ}\mathrm{C}$ (3σ) from -55 to $200\,^{\circ}\mathrm{C}$ after a one-point trim, while drawing only 22 μA. It achieves a resolution FoM of 59 pJ $^{\circ}\mathrm{C}^2$, which is much higher than that of the second prototype, mainly due to the modifications necessary for high temperature operation.

4.3.1 Analog Front-End

Recall from Chap. 1, the base-emitter voltage V_{BE} of a PNP can be expressed as:

$$V_{\mathrm{BE}} = \frac{kT}{q} \ln\left(\frac{I_C}{I_S} + 1\right), \tag{4.4}$$

where k is the Boltzmann constant, q is the electron charge, T is the temperature in Kelvin, I_C is the collector current, and I_S is the saturation current. The base-emitter voltage difference of two identical PNPs biased at a $1{:}p$ collector current ratio can then be expressed as:

$$\Delta V_{\mathrm{BE}} = V_{\mathrm{BE2}} - V_{\mathrm{BE1}} = \frac{kT}{q} \ln\left(\frac{p \cdot I_C + I_S}{I_C + I_S}\right). \tag{4.5}$$

When operation at very high temperatures is targeted, the temperature dependence of the saturation current I_S must be taken into account. The saturation current I_S is given by:

$$I_S(T) = CT^{\eta} \exp\left(\frac{-qV_{g0}}{kT}\right), \tag{4.6}$$

where C and η are constants, while V_{g0} is the extrapolated bandgap voltage at 0 K. The constant $C \propto A_E$, where A_E is the BJT's emitter area, while $\eta \approx 4$, implying that I_S is strongly temperature dependent. The sensitivities of V_{BE} and ΔV_{BE} to temperature variations can be determined from the derivatives:

$$S_{V_{\mathrm{BE}}}^T = \frac{\partial V_{\mathrm{BE}}}{\partial T}, S_{\Delta V_{\mathrm{BE}}}^T = \frac{\partial \Delta V_{\mathrm{BE}}}{\partial T}. \tag{4.7}$$

Figure 4.24 shows the simulated temperature dependency of $S_{V_{\mathrm{BE}}}^T$ and $S_{\Delta V_{\mathrm{BE}}}^T$ for various emitter areas and for a PTAT bias current $I_b = 90\,\mathrm{nA}$ (at $25\,^{\circ}\mathrm{C}$) as in [10]. For $I_S \ll I_C$, $S_{V_{\mathrm{BE}}}^T$ and $S_{\Delta V_{\mathrm{BE}}}^T$ are near-constant, monotonic functions of

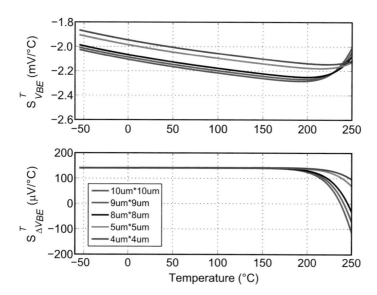

Fig. 4.24 $S^T_{V_{BE}}$ and $S^T_{\Delta V_{BE}}$ versus temperature for different emitter area A_E and for the bias current $I_b = 90\,\text{nA}$

temperature. At high temperatures, however, this condition no longer holds, leading to an inflection point in $S^T_{V_{BE}}$ and $S^T_{\Delta V_{BE}}$. This means that the spread of I_S, will introduce extra temperature-sensing errors that cannot be simply corrected by a one-point trim. Since $C \propto A_E$, the inflection point shifts to higher temperatures as A_E decreases. Alternatively, I_C can be boosted in order to reduce the impact of increases in I_S.

The sensor's analog front-end is shown in Fig. 4.25. Two PNPs, biased at a 5:1 current ratio, generate an accurate bias current I_b, which is then used to bias the PNPs ($A_E = 4\,\mu\text{m} \times 4\,\mu\text{m}$) of the bipolar core. In order to investigate the effect of bias current on sensor accuracy [23], three programmable bias currents can be generated by appropriately connecting the bias resistors R_b. A set of three resistors (each equal to $R_b/5$) in series with the base of Q_{BL} are used to implement the β_F-compensation, as in [3].

As in [10], the impact of the opamp's offset V_{OS} is reduced by chopping it twice per conversion. In [10], a positive feedback opamp topology was used. At high temperatures, however, leakage currents severely impact its gain. In this work, a robust folded-cascode topology is used. Moreover, DEM is applied to mitigate the mismatch of the current sources and PNPs. At high temperatures, the leakage current and the R_{off} of the DEM switches will also modify p [see Eq. (4.5)], leading to extra error in ΔV_{BE} [28]. To mitigate this, the DEM switches are realized with thick oxide NMOS devices and sized appropriately.

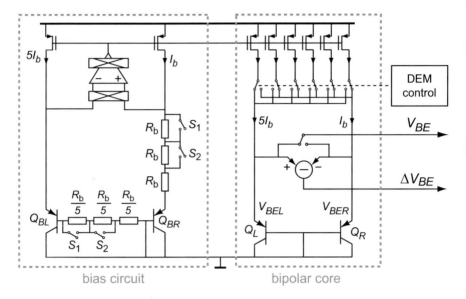

Fig. 4.25 Circuit diagram of the sensor front-end

4.3.2 ADC Design

As in [10], an energy-efficient 2nd-order zoom-ADC was used to digitize the ratio $X = V_{BE}/\Delta V_{BE}$. As previously discussed, since the fine $\Delta\Sigma$-ADC only processes the quantization error of the coarse phase, the output swing of the two integrators is quite small: less than $\pm 3 \cdot \Delta V_{BE}$ ($\pm 125\,\text{mV}$ at $25\,°\text{C}$). In [10], therefore, a telescopic opamp was used in the first integrator while the second integrator used a pair of inverter-based OTAs. At $200\,°\text{C}$, however, the swing increases to about $\pm 200\,\text{mV}$. Moreover, leakage currents then reduce the gain of the inverter-based OTAs significantly. In this work, therefore, both integrators were implemented as folded-cascode OTAs, drawing $1\,\mu\text{A}$ each.

Figure 4.26a shows a pair of unit elements of the capacitor DAC that are used in the sampling network of the ADC, described in Sect. 4.2.3.1 (see Fig. 4.15). Each of the unit capacitors is connected to the bipolar core via two switches S_1 and S_2. Each pair of capacitors samples either V_{BE} (S_1 on) or ΔV_{BE} (S_2 on) or none (both S_1 and S_2 off), depending on the conversion state. At temperatures above $\approx 120\,°\text{C}$, X is less than 6, implying that most of the DAC switches are in the off-state. At these temperatures, however, the leakage current of these switches impacts the current ratio p, while their lower off-resistances causes charge integration errors. This is mitigated by using thick oxide transistors and modifying the switching scheme as shown in Fig. 4.26b. When S_1 and S_2 are in the off-state, an extra switch S_3 ties the nodes A and B to an auxiliary base-emitter voltage $V_{BE,aux}$. This configuration will decrease the voltage across the off switches to either ($V_{BER} - V_{BEaux}$) or

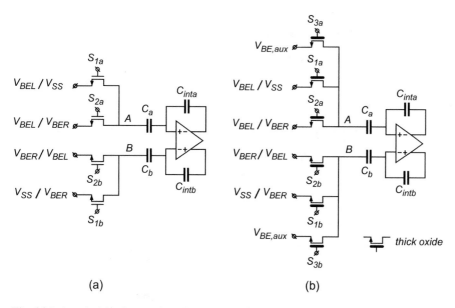

Fig. 4.26 A typical implementation of unit elements in a capacitor DAC (**a**), and the modified switching scheme to cope with the leakage currents and R_{off} at high temperatures (**b**)

($V_{BEL} - V_{BEaux}$), which is small, and thereby significantly reducing the leakage currents through the off switches. It also prevent the nodes A and B from floating and absorbing any leakage currents.

4.3.3   Measurement Results

The 0.1 mm^2 (active area) sensor was realized in a 0.16 μm CMOS-SOI process (Fig. 4.27), in a standard CMOS island, i.e., no SOI-specific features were used. For flexibility, the digital back-end, the control logic, and the $\Delta\Sigma$-ADC's sinc^2 decimation filter were implemented off-chip. Sixteen devices in ceramic DIL packages were characterized from −55 to 200 °C. The sensor's accuracy was measured at three different values of I_b: 330 nA, 660 nA, and 1 μA. Moreover, either Q_R or Q_L, which are biased at I_b and $5 \times I_b$, respectively, can be used to generate V_{BE} (Fig. 4.25). The emitter current I_E at which V_{BE} is generated can, therefore, be programmed between 330 nA (min) and 5 μA (max). As expected, when V_{BE} is generated by $I_E = 5$ μA, the impact of saturation and leakage currents is minimal, leading to an inaccuracy of ±0.4 °C (3σ) after a one-point trim at 30 °C, as shown in Fig. 4.28. The inaccuracy at high temperatures gets worse as the current generating V_{BE} decreases. For a current of $I_E = 330$ nA, the sensor's inaccuracy increases rapidly above 160 °C, and is ±0.6 °C (3σ) at 200 °C, as shown in Fig. 4.28. The 180 and 200 °C temperature points were measured in another setup

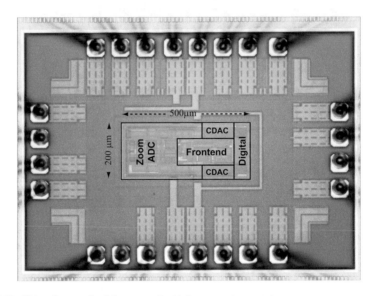

Fig. 4.27 Chip micrograph of the sensor for high temperature sensing

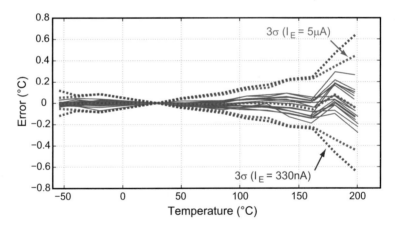

Fig. 4.28 Temperature error of 16 devices after trimming at 30 °C

with a different climate chamber, which may explain the discontinuity between the two sets of measurements. At 25 °C, the sensor draws 22 μA from a 1.6 to 2 V supply, and achieves a kT/C limited resolution of 20 mK (rms) in a conversion time of 4.2 ms. The sensor's performance is summarized in Table 4.5 and compared to that of other sensors capable of operating at high temperatures. Its inaccuracy is comparable to that of state-of-the-art TD sensor in [27], while dissipating much less power.

Table 4.5 Performance summary and comparison with previous work

Parameter	This work ESSCIRC'13 [22]	ISSCC'12 [27]	LM95172 [26]	ADT7312 [25]
Chip area	0.1 mm²	1 mm²	–	–
Supply current (RT)[a]	22 μA	520 μA	500 μA	245 μA
Inaccuracy (trim points)	±0.4 °C (3σ) (1)	±0.4 °C (3σ) (1)	±0.4 °C (trimmed)	±1 °C (trimmed)
Temperature range	−55 to 200 °C	−70 to 200 °C	−40 to 200 °C	−55 to 175 °C
Resolution (T_{conv})	0.02 °C (4.2 ms)	0.075 °C (1.4 s)	0.007 °C (350 ms)	0.007 °C (240 ms)

[a]Excluding the off-chip digital

4.4 Conclusions

In this chapter, various BJT-based sensor prototypes based on zoom-ADCs were demonstrated. By combining the benefits of SAR and $\Delta\Sigma$-ADCs, an accurate, energy-efficient temperature to digital conversion was obtained. By employing a 1st-order zoom-ADC in the first prototype, a low-power sensor was realized, drawing only 4.6 μA. Since the ADC's fine conversion step was based on a slow, 1st-order $\Delta\Sigma$ modulator, it required 100 ms to achieve 15 mK resolution, limiting its energy efficiency. Moreover, a less power-efficient folded-cascode opamp was used to accommodate the large swing and gain requirements of the loop filter. Employing a thermal calibration and a single α-trim at 25 °C, the sensor achieved an accuracy of ±0.2 °C (3σ) over the temperature range from −30 to 125 °C.

To improve the sensor's energy efficiency, a second prototype was realized which achieves similar resolution in about 16× less conversion time, while drawing 25% less supply current. This was achieved by using a 2nd-order zoom-ADC, combined with a new charge-balancing scheme. Simultaneous sampling of V_{BE} and ΔV_{BE} also allowed the use of low-swing, and therefore low-power amplifiers, thus further improving the sensor's energy efficiency. By using a 2nd-order modulator, the required loop-gain to maintain linearity between different fine conversion segments was readily achieved. The sensor's energy efficiency was therefore improved by over 20× compared to the first prototype. Using thermal calibration and digital PTAT trimming at 30 °C, the sensor achieved an accuracy of ±0.15 °C (3σ) over the military temperature range: from −55 to 125 °C. To meet the extreme cost constraints on large volume sensor products, a voltage calibration technique based on electrical measurements was also explored. Compared to thermal calibration, it is significantly faster, requiring only two ADC conversions (200 ms), while achieving comparable accuracy. Moreover, the impact of batch-to-batch spread and plastic packaging on sensor's accuracy was explored. As observed, both batch-to-batch spread and plastic packaging can cause temperature reading shifts in the order of 0.4–0.5 °C over the military temperature range.

Last but not least, a BJT-based temperature sensor for very high temperature sensing was also implemented. To mitigate the impact of leakage and BJT's saturation current at high temperatures, the bias current and the emitter area of BJTs ($A_E = 4\,\mu\text{m} \times 4\,\mu\text{m}$) were carefully optimized and robust circuit-level techniques were used. After a one-point trim at 30 °C, the sensor achieved an accuracy of ± 0.4 °C (3σ) from -55 to 200 °C, which is similar to that of state-of-the-art thermal diffusivity (TD) sensors, but it draws more than an order of magnitude less current.

References

1. K. Souri, K.A.A. Makinwa, A 0.12 mm² 7.4 μW micropower temperature sensor with an inaccuracy of ± 0.2 °C (3σ) from -30 °C to 125 °C. IEEE J. Solid State Circuits, **46**(7), 1693–1700 (2011)
2. M.A.P. Pertijs, J.H. Huijsing, *Precision Temperature Sensors in CMOS Technology* (Springer, Dordrecht, 2006)
3. M.A.P. Pertijs, K.A.A. Makinwa, J.H. Huijsing, A CMOS temperature sensor with a 3σ inaccuracy of ± 0.1 °C from -55 °C to 125 °C. IEEE J. Solid State Circuits **40**(12), 2805–2815 (2005)
4. K. Souri, M. Kashmiri, K.A.A. Makinwa, A CMOS temperature sensor with an energy-efficient zoom ADC and an inaccuracy of ± 0.25 °C (3σ) from -40 to 125 °C, in *Digest of Technical Papers ISSCC*, Feb 2010, pp. 310–311
5. K. Souri, K.A.A. Makinwa, A 0.12 mm² 7.4 μW micropower temperature sensor with an inaccuracy of ± 0.2 °C (3σ) from -30 °C to 125 °C, in *Proceedings of ESSCIRC*, Sept 2010, pp. 282–285
6. M.G. Degrauwe, J. Rijmenants, E.A. Vittoz, H.J. DeMan, Adaptive biasing CMOS amplifiers. IEEE J. Solid State Circuits **SC-17**, 522–528 (1982)
7. D. Schinkel, R.P. de Boer, A.J. Annema, A.J.M. van Tuijl, A 1-V 15 μW high-accuracy temperature switch. Springer Analog Integr. Circ. Sig. Process **41**(1), 13–20 (2004)
8. A.L. Aita, M.A.P. Pertijs, K.A.A. Makinwa, J.H. Huijsing, A CMOS smart temperature sensor with a batch-calibrated inaccuracy of ± 0.25 °C (3σ) from -70 °C to 130 °C, in *Digest of Technical Papers ISSCC*, Feb 2009, pp. 342–343
9. R.C. Yen, P.R. Gray, A MOS switched capacitor instrumentation amplifier. IEEE J. Solid State Circuits **SC-17**(6), 1091–1097 (1982)
10. K. Souri, Y. Chae, K.A.A. Makinwa, A CMOS temperature sensor with a voltage-calibrated inaccuracy of ± 0.15 °C (3σ) from -55 °C to 125 °C. IEEE J. Solid-State Circuits **48**(1), 292–301 (2013)
11. K.Y. Nam, S.-M. Lee, D.K. Su, B.A. Wooley, A low-voltage low-power sigma-delta modulator for broadband analog-to-digital conversion. IEEE J. Solid State Circuits **40**(9), 1855–1864 (2005)
12. Y. Chae, G. Han, Low voltage, low power, inverter-based switched-capacitor delta-sigma modulator. IEEE J. Solid State Circuits **44**(2), 458–472 (2009)
13. J. Markus, J. Silva, G.C. Temes, Theory and applications of incremental delta-sigma converters. IEEE Trans. Circuits Syst. I **51**(4), 678–690 (2004)
14. K. Souri, K.A.A. Makinwa, Ramp calibration of temperature sensors, in *Proceedings of IWASI*, June 2011, pp. 282–285
15. M.A.P. Pertijs, A.L. Aita, K.A.A. Makinwa, J.H. Huijsing, Voltage calibration of smart temperature sensors, in *Proceedings of IEEE Sensors*, Oct 2008, pp. 756–759
16. A. Hastings, *The Art of Analog Layout* (Prentice Hall, New Jersey, 2001)

17. F. Fruett, G.C.M. Meijer, A. Bakker, Minimization of the mechanical-stress-induced inaccuracy in bandgap voltage references. IEEE J. Solid State Circuits **38**(7), 1288–1291 (2003)
18. J.F. Creemer, F. Fruett, G.C. Meijer, P.J. French, The piezojunction effect in silicon sensors and circuits and its relation to piezoresistance. IEEE Sens. J. **1**(2), 98–108 (2001)
19. S.Z. Asl et al., A $1.55 \times 0.85 \, mm^2$ 3 ppm $1.0 \, \mu A$ 32.768 kHz MEMS-based oscillator, in *Digest of Technical Papers*, Feb 2014, pp. 226–227
20. C.-K. Wu et al., A $80 \, kS/s$ $36 \, \mu W$ resistor-based temperature sensor using BGR-free SAR ADC with an unevenly-weighted resistor string in $0.18 \, \mu m$ CMOS, in *IEEE Symposium on VLSI Circuits*, June 2011, pp. 222–223
21. M. Law, A. Bermark, A 405-nW CMOS temperature sensor based on linear MOS operation. IEEE Trans. Circuits Syst. II **56**(12), 891–895 (2009)
22. K. Souri, K. Souri, K.A.A. Makinwa, A $40 \, \mu W$ CMOS temperature sensor with an inaccuracy of $\pm 0.4 \, °C$ (3σ) from $-55 \, °C$ to $200 \, °C$, in *Proceedings of ESSCIRC*, Sept 2013, pp. 221–224
23. A.L. Aita et al., A low-power CMOS smart temperature sensor with a batch-calibrated inaccuracy of $\pm 0.25 \, °C$ (3σ) from -70 to $130 \, °C$. IEEE Sens. J. **13**, 1840–1848 (2013)
24. R.A. Bianchi et al., CMOS compatible temperature sensor based on the lateral bipolar transistor for very wide temperature range applications. Sens. Actuators A **71**(1–2), 3–9 (1998)
25. ADT7312 Data Sheet (Analog Devices Inc., Norwood, 2012), www.analog.com
26. LM95172 Data Sheet (Texas Instruments Inc., Dallas, 2013), www.ti.com
27. C.P.L. van Vroonhoven, K.A.A. Makinwa, A $\pm 0.4 \, °C$ (3σ) -70 to $200 \, °C$ time-domain temperature sensor based on heat diffusion in Si and SiO2, in *Digest of Technical Papers*, Feb 2012, pp. 204–206
28. A.L. Aita, K.A.A. Makinwa, Low-power operation of a precision CMOS temperature sensor based on substrate PNPs, in *Proceedings of IEEE Sensors*, Oct 2007, pp. 856–859

Chapter 5
All-CMOS Precision Temperature Sensors

In CMOS technology, BJT-based sensors are usually the temperature sensors of choice due to their decent accuracy after a single temperature trimming, e.g., $\pm 0.2\,^\circ\text{C}$ (3σ) over the military temperature range: -55 to $125\,^\circ\text{C}$ [1–3]. They also achieve low supply-sensitivity, typically in the order of $0.1\,^\circ\text{C/V}$. However, BJT-based sensors typically require supply voltages above 1 V, since V_{BE} will be about 0.8 V at $-55\,^\circ\text{C}$ and some headroom is required for the current source (often cascoded) that biases the BJT. This restricts the use of such sensors in battery-powered systems, and also restricts the temperature range of implementations in nanometer CMOS [4]. However, from an energy efficiency perspective, a lower V_{DD} value is preferred.

To achieve sub-1V operation, MOSFET-based temperature sensors have been proposed [5]. When biased in the sub-threshold region, the drain current I_D and the gate-source voltage V_{GS} of a MOSFET exhibit a temperature-dependent exponential relationship, similar to that between the collector current I_C and V_{BE} of a BJT. As a result, MOSFETs can also be used as temperature sensing elements. At $-55\,^\circ\text{C}$, however, V_{GS} will only be a few hundred millivolts, which makes sub-1V operation possible. Unfortunately, process spread affects two parameters of a MOSFET: the threshold voltage V_T and the charge mobility μ. As a result, such sensors exhibit greater inaccuracy than BJT-based sensors, e.g., $-1.8/+1\,^\circ\text{C}$ from 10 to $80\,^\circ\text{C}$ after a single temperature trim [5]. The propagation delay of a CMOS inverter can also be used as a measure of temperature. Compared to BJT-based sensors, however, such sensors suffer from much greater supply sensitivity (about $10\,^\circ\text{C/V}$) and greater inaccuracy ($-0.4/+0.6\,^\circ\text{C}$) over a limited range from 0 to $90\,^\circ\text{C}$, even after a more-costly two-temperature trim [6].

This chapter describes the design of two temperature sensor prototypes based on dynamic threshold MOS transistors (DTMOSTs) [7–9]. By using the DTMOST configuration, i.e., by connecting the body of a MOSFET to its gate, a near-ideal diode characteristic can be realized. The resulting device can then replace the BJTs used in precision temperature sensors, e.g., sensors described in Chap. 4. In the

© Springer International Publishing AG 2018

K. Souri, K.A.A. Makinwa, *Energy-Efficient Smart Temperature Sensors in CMOS Technology*, Analog Circuits and Signal Processing, DOI 10.1007/978-3-319-62307-8_5

following, the design of a first prototype DTMOST-based sensor is described. After a two-temperature trim, the sensor achieves an inaccuracy of $\pm 0.1\,^\circ$C (3σ) over the military temperature range (-55 to $125\,^\circ$C). This represents a $5\times$ improvement in accuracy over previously reported MOSFET-based temperature sensors. Employing a one-point trim increases the inaccuracy to $\pm 0.4\,^\circ$C (3σ), which is only a factor $2\times$ worse compared to, similarly trimmed, BJT-based sensors. In a second prototype, the low-voltage capability of DTMOSTs is then exploited to realize a sub-1V, sub-μW precision sensor.

5.1   DTMOSTs as Sensing Element [10]

A DTMOST is a standard MOSFET whose body and gate terminals are connected together. This causes its threshold voltage to vary dynamically, hence the name [7]. Compared to a diode-connected MOSFET, the V_{GS}–I_D characteristic of a diode-connected DTMOST, i.e., a DTMOST diode, is less sensitive to V_T spread [7–9]. In this section, a first DTMOST-based sensor prototype will be presented, which after a single temperature trim achieves an inaccuracy of $\pm 0.4\,^\circ$C (3σ) over the military temperature range, and improves to $\pm 0.1\,^\circ$C (3σ) after a two-temperature trim. Compared to state-of-the-art MOSFET-based sensors [5, 6], this represents a $5\times$ improvement in accuracy.

5.1.1   Operating Principle

The use of a DTMOST as a temperature sensing element was described in Chap. 1. Like the base-emitter voltage V_{BE} of a BJT, the gate-source voltage V_{GS} of a DTMOST diode exhibits complementary-to-absolute temperature (CTAT) behavior, but with a smaller temperature coefficient: $\approx -1\,$mV/$^\circ$C. In a similar manner, the difference in the gate-source voltage ΔV_{GS} of two DTMOST diodes is PTAT: $\Delta V_{GS} = (kT/q) \cdot \ln(p)$, where p is the bias current ratio and kT/q is the thermal voltage. Furthermore, a DTMOST has an effective bandgap voltage of roughly 0.6 V [7], i.e., about half the bandgap voltage of a BJT. This suggests that DTMOSTs can be used to implement sub-1V bandgap voltage references [7, 11] and temperature sensors in standard CMOS. Since DTMOSTs are just alternatively connected MOSFETs, such circuits can be designed and optimized with the help of standard compact models. This is another advantage over parasitic BJTs, which require other models, and which are often not very accurately characterized. Last but not least, the gate-source voltage V_{GS} of a DTMOST exhibits less spread, about half that of a normal MOSFET [7, 8], which is a promising feature to realize accurate temperature sensors.

5.1.2   Temperature Sensor Design

Figure 5.1 shows the CTAT and PTAT behavior of the V_{GS} and ΔV_{GS} voltages of a DTMOST diode. As in a bandgap voltage reference, a reference voltage can be derived from a linear combination of ΔV_{GS} and V_{GS} [7]: $V_{REF} = V_{GS} + \alpha \cdot \Delta V_{GS}$, where α is the fixed gain factor required to obtain a nominally zero temperature coefficient. As in a bandgap (BJT-based) temperature sensor [1], an analog-to-digital converter (ADC) can then digitize the PTAT ratio μ between the PTAT voltage $\alpha \cdot \Delta V_{GS}$ and V_{REF}, to obtain:

$$\mu = \frac{\alpha \cdot \Delta V_{GS}}{V_{GS} + \alpha \cdot \Delta V_{GS}}, \tag{5.1}$$

where μ varies between 0 and 1. For the bias current ratio $p = 5$ used in this design, $\alpha \approx 8$. A linear scaling of μ then results in a digital output in degrees Celsius: $D_{out} = A \cdot \mu + B$, where $A \approx 600\,K$, and $B \approx -273\,K$, as presented in Eq. (2.9).

As shown in the previous chapter, the ratio $X = V_{GS}/\Delta V_{GS}$ can now be used as a measure of temperature. For $p = 5$, the resulting nonlinear ratio $X = V_{GS}/\Delta V_{GS}$ ranges between 3 and 12 for temperatures ranging from -55 to $125\,°C$. This ratio can then be digitized by an ADC, and the PTAT ratio μ is readily derived as: $\mu = \alpha/(\alpha + X)$ (Fig. 5.2). At $-55\,°C$ the V_{GS} of a DTMOST diode is about 400 mV, about half that of a diode-connected BJT.

In the front-end of the proposed sensor (Fig. 5.1), the voltages V_{GS} and ΔV_{GS} are generated by two identical DTMOST diodes ($W/L = 80\,\mu m/0.65\,\mu m$) biased at $I_b = 90\,nA$ (at $25\,°C$) and $5 \cdot I_b$. The required current ratio $p = 5$ is established with the help of six unit current sources. As in [1, 3], dynamic element matching (DEM) is used to average out the mismatch between these current sources and between the two DTMOST diodes, as this would otherwise cause significant temperature-sensing errors. The biasing currents are generated by the same PNP-based bias circuit used in the BJT-based sensor of [3]. This choice allows a fair comparison to be made between BJT- and DTMOST-based temperature sensors, but also precludes sub-1V operation.

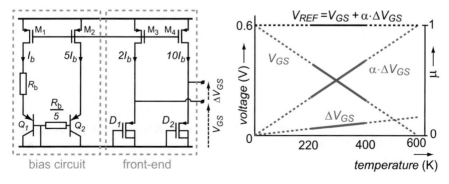

Fig. 5.1 Temperature dependence of V_{GS} and ΔV_{GS}, generated by two DTMOST diodes biased in sub-threshold region

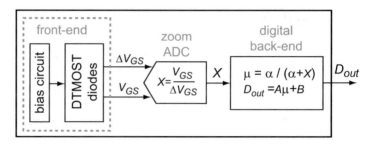

Fig. 5.2 Block diagram of the smart temperature sensor

Another source of current-ratio error, especially at high temperatures, is the source-to-bulk leakage current of the DTMOST diodes. This was minimized by appropriate device sizing using standard device models [9]. The area of the optimized devices is then in the same order as that of the PNPs in [3]. It should be noted that this nonideal leakage current is two orders of magnitude less than, the similarly nonideal, base current of a parasitic PNP transistor in the same process. As a result, DTMOST-based sensors, unlike their PNP-based counterparts, do not require extra β-compensation-like circuitry [1, 3].

The ratio $X = V_{GS}/\Delta V_{GS}$ is accurately digitized by a 16-bit zoom-ADC, which combines the advantages of a SAR-ADC and a 1st-order $\Delta\Sigma$ converter in a two-step conversion scheme [3]. As described in Sect. 3.3, a SAR algorithm is first employed to find the integer part of X, denoted by n, by comparing V_{GS} to integer multiples of ΔV_{GS}. The fractional part μ' is then determined by a 1st-order $\Delta\Sigma$-ADC, whose references are chosen so as to zoom into the region determined by the SAR algorithm, i.e., from $n \cdot \Delta V_{GS}$ to $(n + 1) \cdot \Delta V_{GS}$. A modified 1st-order SC $\Delta\Sigma$-ADC is used to implement both conversion steps, resulting in a compact, energy-efficient ADC.

5.1.3 Measurement Results

The temperature sensor was realized in a standard $0.16\,\mu m$ CMOS process with five metal layers (Fig. 5.3). The chip has an active area of $0.12\,mm^2$ and consumes $8.6\,\mu W$ from a 1.8 V supply at 30 °C. The sensor has a supply sensitivity of 0.1 °C/V over a supply range of 1.6–2 V. For flexibility, the digital back-end, the control logic, and the $\Delta\Sigma$ modulator's $sinc^2$ decimation filter were implemented off-chip.

Twenty devices from one batch were packaged in ceramic DIL packages, placed in a climate chamber and then characterized over the military temperature range: −55 to 125 °C. As shown in Fig. 5.4, the sensor's batch-calibrated inaccuracy was about ±1.5 °C (3σ), with a PTAT-like characteristic and a 3rd-order systematic nonlinearity of about ±0.1 °C. After digital nonlinearity compensation and a single

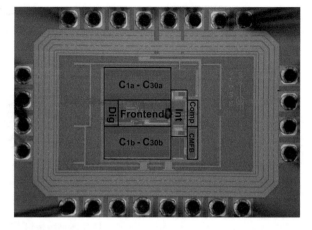

Fig. 5.3 Chip micrograph of the first prototype DTMOST-based sensor employing a 1st-order zoom-ADC

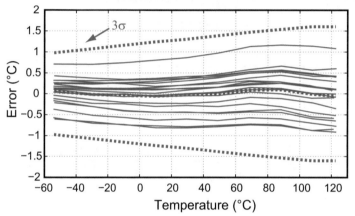

Fig. 5.4 Temperature error of 20 measured DTMOST-based sensors before trimming (linear fit)

offset-trim at 30 °C, the sensor's inaccuracy was reduced to ±0.4 °C (3σ), as shown in Fig. 5.5. An α-trim [3] resulted in slightly worse inaccuracy: ±0.5 °C (3σ).

As shown in Fig. 5.6, a two-temperature trim (commonly used by MOSFET-based temperature sensors [6]) at −10 and 90 °C, reduces the sensor's inaccuracy to ±0.1 °C. At 5 conversions/s (1024 $\Delta\Sigma$-cycles), the sensor achieves a resolution of 33 mK (rms), which is lower than that of BJT-based sensor in [3], which uses a similar readout topology. As previously discussed, for a similar 5:1 ratio of bias currents used in the front-end, the ratio X in a DTMOST-based sensor is about half that of a BJT-based sensor. The number of ΔV_{GS} sampling capacitors are, therefore, proportionally smaller and results in a larger kT/C noise. The sensor's performance is summarized in Table 5.1. Compared to the prior-art MOSFET-based temperature sensors [5, 6], this represents 5× more accuracy over a much wider temperature range and proves DTMOSTs as a promising sensing element.

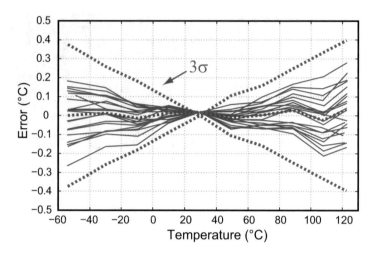

Fig. 5.5 Temperature error of 20 measured DTMOST-based sensors after a single offset trim at 30 °C

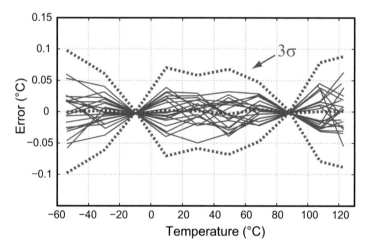

Fig. 5.6 Temperature error of 20 measured DTMOST-based sensors after two-temperature trim at −10 and 90 °C

For comparison, a second chip was realized with essentially the same readout circuitry, but in which the DTMOST diodes were replaced by PNPs. For $p = 5$, however, the ratio $X = V_{BE}/\Delta V_{BE}$ then spans a larger range: varying from 6 to 28 from −55 to 125 °C [3]. To cover this range, the range of the zoom-ADC's SAR phase was extended, by proportionally increasing the number of elements in its capacitor DAC. As shown in Fig. 5.7, the inaccuracy of the PNP-based sensor is ±0.2 °C (3σ) after a single α-trim: only 2× less than that of, a similarly trimmed, DTMOST-based sensor.

Table 5.1 Performance summary and comparison with previous work

Parameter	This work ESSCIRC'11 [10]	S&A'11 [5]	JSSC'10 [6]
CMOS technology	0.16 μm	0.35 μm	0.35 μm
Chip area	0.12 mm²	0.08 mm²	0.06 mm²
Supply current (RT)[a]	4.7 μA	4.5 μA	10.6 mA
Supply sensitivity	0.1 °C/V	2.5 °C/V (at 30 °C)	10 °C/V (at 30 °C)
Inaccuracy (calibration points)	±0.1 °C(3σ), (2) ±0.4 °C(3σ), (1)	−1.8 °C/ + 1 °C(max), (1)	−0.4 °C/ + 0.6 °C(max), (2)
Temperature range	−55 to 125 °C	10 to 80 °C	0 to 90 °C
Resolution (T_{conv})	0.033 °C (200 ms)	0.1 °C (10 ms)	0.0918 °C (90 μs)

[a] Excluding the off-chip digital

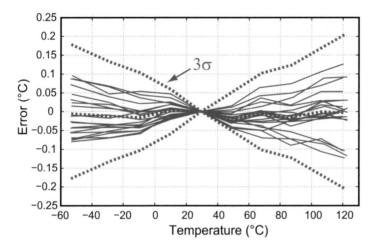

Fig. 5.7 Temperature error of 20 measured PNP-based sensors after a single α-trim at 30 °C

5.2 A Sub-1V All-CMOS Temperature Sensor [12]

The first DTMOST-based prototype presented in the previous section is not capable of operating at sub-1V supply voltages, due to the use of a BJT-based bias circuit. In this section a second prototype will be presented. It is an ultra-low-power all-CMOS temperature sensor, which operates from a 0.85 V supply and draws only 700 nA. To date, this is the only all-CMOS temperature sensor to achieve both sub-1V operation and a high accuracy of ±0.4 °C (3σ) over a wide temperature range (−40 to 125 °C), while employing only a single room-temperature trim.

5.2.1 Sensor Front-End

Consider the bias circuit shown in Fig. 5.1. As already discussed in Chap. 3, the
circuit generates a PTAT, β_F-dependent bias current. It is then mirrored into the
sensor front-end to bias two identical DTMOSTs at a 5:1 current ratio. Unlike BJTs,
however, DTMOSTs do not suffer from an equivalent to base current, thus ruling out
the need for such β_F-compensation circuitry. In this design, therefore, the simple
sensor front-end shown in Fig. 5.8a has been adopted. A pair of DTMOST diodes
with a 1:2 area ratio are biased with identical bias currents $I = \Delta V_{GS}/R_b$, which are
enforced by a self-biased opamp configuration.

Since $V_{GS} \approx 0.4$ V at $-55\,°C$, and assuming some voltage headroom for the
cascoded current sources (not shown), using a supply voltage below 1 V is readily
feasible. The challenge then would be to design a low-power opamp which operates
from a sub-1V supply. A well-known, energy-efficient solution is the current-
voltage mirror (CVM) circuit of Fig. 5.8b [13]. The top PMOS current mirrors M_{P1}
and M_{P2} force $I_X = I_Y$, while M_{N1} and M_{N2} force $V_X = V_Y$, given their equal sizing
and drain current values. Two diode-connected transistors M_{N1} and M_{P2} ensure an
overall negative feedback around the loop. Although extremely compact and energy-
efficient, this circuit exhibits poor regulation. Since the drain-source voltage of the
PMOS current mirrors are different, channel length modulation makes $I_X \neq I_Y$,
thereby resulting in a systematic offset between the V_{GS} voltages of M_{N1} and M_{N2}.
This offset is subject to the supply and temperature changes and will result in an
untrimmable error at the output of the sensor.

To solve the aforementioned issues, the circuit of Fig. 5.9a has been implemented
in this work. The so-called symmetrically matched current-voltage mirror (SM
CVM) [14] forces the PMOS current mirrors to essentially have the same gate, drain,
and source voltages, thereby forcing equal currents $I_X = I_Y$, and in turn, maintain

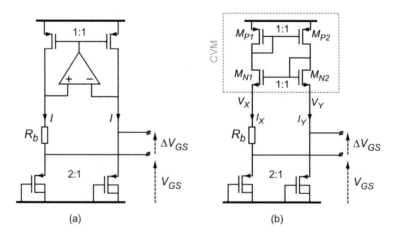

Fig. 5.8 Circuit diagram of the DTMOST-based front-end using an amplifier (**a**), and a current-
voltage mirror (**b**)

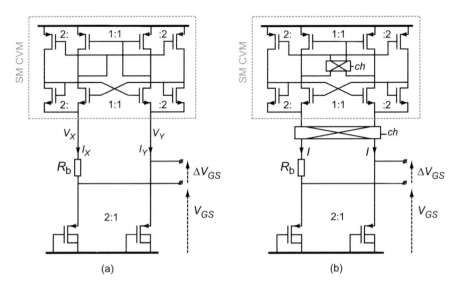

Fig. 5.9 Circuit diagram of the DTMOST-based front-end using a SM CVM (**a**). Applying chopping to improve the accuracy of SM CVM (**b**)

$V_X = V_Y$. The voltage headroom for a CVM to operate is $V_{GS,PMOS} + V_{DS,NMOS}$. Given the low bias currents in each branch (90 nA in this design), the top current mirrors are designed to operate in weak inversion, thus reducing their V_{GS} voltage. Assuming $V_{GS,PMOS} = 0.3$ V and $V_{DS,NMOS} = 100$–150 mV, a voltage headroom of ≈ 450 mV is sufficient, thus enabling sub-1V operation.

5.2.2 Accuracy Issues

A main source of error is the CVM's input-referred offset and $1/f$ noise, which directly add to ΔV_{GS} and thus impact the accuracy of the bias currents, and hence of both V_{GS} and ΔV_{GS}. Such errors are mitigated by incorporating choppers into the CVM circuitry as shown in Fig. 5.9b. While the input chopper swaps the two inputs of the amplifier, thereby modulating the offset and $1/f$ noise to chopping frequency f_{chop}, a second chopper inside the CVM sustains the negative feedback polarity. The succeeding $\Delta\Sigma$ modulator and low-pass digital decimation filter then filter out the high-frequency components present at f_{chop}.

To minimize the effect of mismatch between the DTMOSTs, which would otherwise impact the accuracy of ΔV_{GS}, the 1:2 area ratio is established by incorporating three unit DTMOSTs ($W/L = 90\,\mu\text{m}/0.7\,\mu\text{m}$ for each) in a dynamic element matching (DEM) scheme, as shown in Fig. 5.10. Since the associated DEM switches carry bias current, Kelvin connections are used to accurately read out V_{GS} and ΔV_{GS}. Unlike in [3] where off-chip DEM control was used, here, an on-chip DEM logic based on a 3-bit ring counter is employed.

Fig. 5.10 DEM and Kelvin
connection used to improve
the accuracy of the front-end

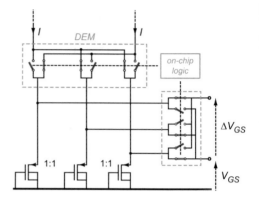

5.2.3 System Diagram

The sensor's block diagram is shown in Fig. 5.11a. It consists of the DTMOST front-end of Fig. 5.9b, a 2nd-order inverter-based zoom-ADC, a voltage doubler and some control logic. As in [3], the zoom-ADC uses a power-efficient coarse/fine algorithm to convert the front-end's output voltages V_{GS} and ΔV_{GS} into a temperature-dependent ratio $X = V_{GS}/\Delta V_{GS}$. For the choice of DTMOSTs with a 1:2 area ratio, X varies from 5 to 28 over the temperature range from -40 to $125\,^{\circ}\mathrm{C}$. As before, an off-chip digital back-end then computes a PTAT ratio $\mu = \alpha/(\alpha + X)$, where α is a gain factor, which could be also used as a trimming knob to compensate for V_{GS} spread.

Figure 5.11b shows the timing diagram of a full temperature-to-digital conversion. As shown, each conversion starts with a 5-bit SAR, which determines the integer part of X, i.e., n. It is then followed with a fine $\Delta\Sigma$ conversion of length $N_{\Delta\Sigma}$ to determine the fraction μ'. Due to mismatch, the X value may change significantly during the two states of the chopped front-end, thus resulting in different coarse values n. This issue is solved as follows: first the chopping state is set to "0" and a coarse/fine conversion is performed, resulting in $X_1 = n_1 + \mu'_1$. The chopping state is then swapped and a second coarse/fine conversion is done to obtain $X_2 = n_2 + \mu'_2$. The final result is achieved by averaging the results over two conversions, i.e., $X = (X_1 + X_2)/2$. To minimize the ADC's own residual offset, system-level chopping with similar timing is employed, as shown in Fig. 5.11b.

5.2.4 Power Domains

The sensor has two supply voltages: an analog supply AVDD, which powers the front-end and the ADC, and a digital supply DVDD, which powers the voltage doubler. AVDD can range between 0.85 and 1.2 V, while DVDD = 0.9 V. The voltage doubler provides the drive logic block with ≈ 1.8 V that, in turn, drives the gates

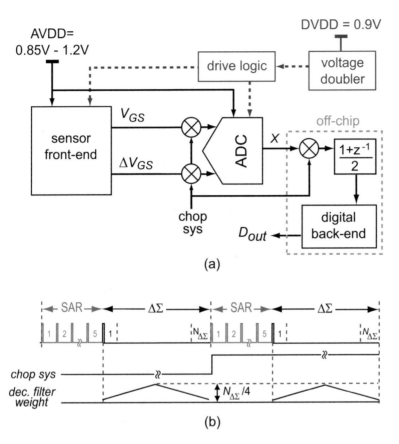

(a)

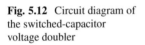

(b)

Fig. 5.11 *Top*, (**a**): The sensor's block diagram. *Bottom*, (**b**): Timing diagram of a temperature conversion

Fig. 5.12 Circuit diagram of the switched-capacitor voltage doubler

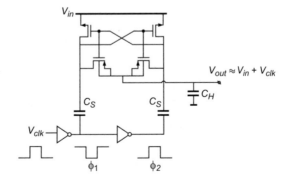

of NMOS switches that sample V_{GS} and ΔV_{GS}, thus facilitating the use of sub-1V supply voltages. The voltage doubler employs a charge-pump topology as in Fig. 5.12. Running at 80 kHz, the two anti-phase clocks ϕ_1, ϕ_2 with the amplitude of V_{clk}, along with the input voltage V_{in} are used to periodically charge the sampling capacitors C_S, and thus delivering an output voltage $V_{out} \approx V_{clk} + V_{in}$. The holding capacitor $C_H = 1\,pF$ is used to reduce the output ripples to less than $30\,mV$. Since the power dissipated in the drive logic is small, the area/power overhead due to the use of the charge-pump block is minimal.

5.2.5 Inverter-Based Zoom ADC

The heart of the zoom-ADC is a feed-forward 2nd-order switched-capacitor $\Delta\Sigma$-ADC, as shown in Fig. 5.13. At its input, a cap-DAC with 30-unit elements (each $60\,fF$) samples either V_{GS} or $k \cdot \Delta V_{GS}$, where $k = [1 \ldots 30]$. The two capacitors $C_G = 0.5 \cdot C_{Sk}$ are used to implement guard-banding.

In contrast to [3], both integrators are built around pseudo-differential inverter-based amplifiers, thus fully exploiting the reduced integrator swing conferred by zooming. The first integrator draws $135\,nA$ while the second, less critical, integrator draws only $66\,nA$. Figure 5.14 shows the implementation details of the 1st integrator's OTA. The 2nd integrator essentially uses the same topology, except that its current is scaled by half. To decrease the inverter's sensitivity to power supply and process spread, a dynamic current-biasing technique is used. During the auto-zeroing phase ϕ_1, M_{N1}, and M_{P1} are diode-connected and biased with two current sources. The left plates of the offset storing capacitors C_{OS} are connected to the signal ground in this phase, thus storing the operating bias voltages of M_{N1} and M_{P1}. The bias voltages V_{b1} and V_{b2} are chosen such that M_{N2}, M_{P2} are essentially

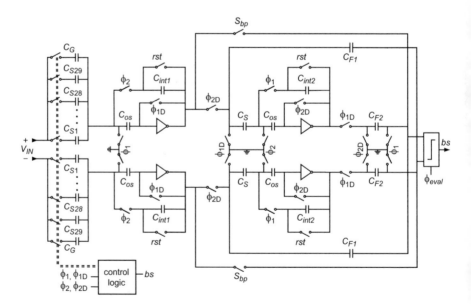

Fig. 5.13 Circuit diagram of the 2nd-order inverter-based zoom-ADC

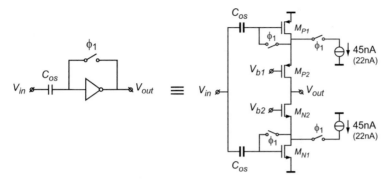

Fig. 5.14 Circuit diagram of the inverter-based OTA. A PTAT bias current of 45 and 22 nA (at room temperature) is used in the first and second OTAs, respectively

off during ϕ_1. The two bias currents are mirrored from the front-end's precision bias circuit to ensure robustness to supply and process variations. In ϕ_2, the two biasing current sources are disconnected and M_{N1} and M_{P1} are configured as common-source amplifiers and together form a class-AB amplifier, with a virtual ground at V_{in}. Since the output voltage swing requirement is reduced to about ± 200 mV by the prior coarse conversion step, cascoding of M_{N1} and M_{P1} is readily possible, thus enhancing the inverter's output resistance, and hence its DC gain. A passive summation network at the input of quantizer combines the output of 2nd integrator with that of the 1st integrator via the feed-forward capacitor C_{F1}, as shown in Fig. 5.13. During the coarse conversion, the first integrator computes $V_{GS} - k \cdot \Delta V_{GS}$, while its output is connected directly to the comparator via the switch S_{bp}. Off-chip logic then implements the SAR algorithm by applying trial values k to the chip and monitoring the comparator's output.

During the fine conversion, the mismatch between the unit elements of the cap-DAC is mitigated by the use of DEM. In contrast to [3], the required DEM logic is implemented on-chip. Figure 5.15 shows the circuit diagram of the implemented on-chip DEM logic. It consists of a 30-bit circular shift-register (SR), with an effective length of $n + 3$ bits, where n is the result of the coarse conversion. The SR length is defined by a periodic reset signal, which is generated by means of an on-chip 5-bit down-counter, that is preloaded with $n + 3$ at every falling edge of the resulting reset signal. Resetting the SR loads it with a single logic "1" appearing at m_0 output, which then circulates on every succeeding clock pulse. This bit (via the m_i outputs) is then used to select the single capacitor that samples V_{GS}.

During the fine conversion, either n or $n + 2$ capacitors will be used to sample ΔV_{GS}, depending on the bitstream output bs. An on-chip decoder is used to convert the binary code $n + 3$ into a thermometer code. The resulting thermometer code $[k_0 k_1 \ldots k_{29}]$, the V_{GS} sampling control $[m_0 m_1 \ldots m_{29}]$, and the bitstream output bs are then applied to a simple combinational logic to generate the ΔV_{GS} sampling control $[t_0 t_1 \ldots t_{29}]$. During the coarse conversion, the SR is reset, so that the same capacitors are always used for the SAR conversion.

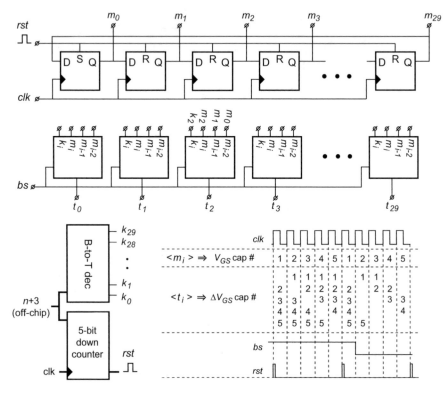

Fig. 5.15 *Top*: On-chip DEM logic for the zoom-ADC's Cap-DAC. *Bottom*: Timing diagram for a 5-element Cap-DAC as an example

5.2.6 Prototype and Measurement Results

The second prototype was also realized in a standard 0.16 μm CMOS process. Figure 5.16 shows the chip micrograph of the sensor. It occupies 0.085 mm² and draws 700 nA from a 0.85 V supply. The front-end and ADC draw 560 nA, while the voltage doubler and the rest of the on-chip digital circuitry draw 140 nA. For flexibility, the SAR logic and the sinc² decimation filter were implemented off-chip. However, simulations show that implementing them on-chip would only incur an extra 10 nW per conversion. With DVDD fixed at 0.9 V, AVDD was varied from 0.85 to 1.2 V. The corresponding supply sensitivity of the front-end and ADC was 0.45 °C/V.

A total of 16 devices in ceramic DIL packages were characterized over the temperature range from −40 to 125 °C. As shown in Fig. 5.17 (top), their batch-calibrated inaccuracy was ±1 °C (3σ, 16 devices), with a residual curvature of only 0.03 °C. Compared to the first prototype (Fig. 5.4), this design exhibits better untrimmed accuracy, but with a CTAT rather than PTAT profile. At cold, the bias

Fig. 5.16 Chip micrograph
of the second prototype
DTMOST-based sensor
employing a 2nd-order
zoom-ADC

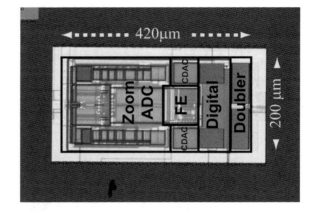

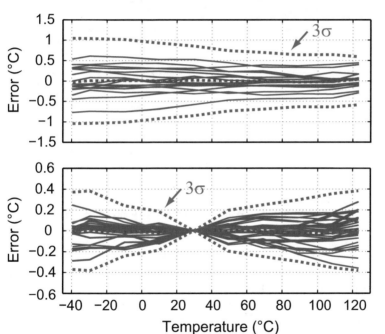

Fig. 5.17 Measured temperature error of 16 samples before trimming (*top*) and after a single α-trim at 30 °C (*bottom*)

current in the sensor front-end is minimum, while the number of sampling capacitors
is maximum. Therefore, any settling errors will be larger at cold, which could
explain the CTAT profile of the untrimmed inaccuracy shown in Fig. 5.17 (top).
Furthermore, the residual offset of the SM CVM will appear in series with ΔV_{GS},
and thus directly impacts the sensor's accuracy. This effect is more severe at cold,
when ΔV_{GS} is smaller. After an α-trim at 30 °C, the inaccuracy improves to ± 0.4 °C

Table 5.2 Performance summary and comparison with previous work

Parameter	Second prototype ISSCC'14 [12]	First prototype ESSCIRC'11 [10]	JSSC'10 [16]	CICC'08 [17]	VLSI'11 [18]
Sensor type	DTMOST	DTMOST	MOSFET	MOSFET	Resistor
CMOS technology	$0.16\,\mu\text{m}$	$0.16\,\mu\text{m}$	$0.18\,\mu\text{m}$	$0.18\,\mu\text{m}$	$0.18\,\mu\text{m}$
Chip area	$0.085\,\text{mm}^2$	$0.12\,\text{mm}^2$	$0.142\,\text{mm}^2$	$0.05\,\text{mm}^2$	$0.18\,\text{mm}^2$
Supply current (RT)[a]	700 nA	$4.6\,\mu\text{A}$	190 nA	220 nA	$20\,\mu\text{A}$
Supply voltage	0.85–1.2 V	1.5–2 V	0.5 V (sensor) 1.0 V (digital)	1 V	1.2–2 V
Supply sensitivity	$0.45\,^\circ\text{C/V}$	$0.1\,^\circ\text{C/V}$	–	Supply referenced	$0.625\,^\circ\text{C/V}$
Inaccuracy (calibration points)	$\pm 0.15\,^\circ\text{C}(3\sigma)$ (1)	$\pm 0.2\,^\circ\text{C}(3\sigma)$ (1)	$-0.8\,^\circ\text{C/+}$ $1\,^\circ\text{C}$ (2)	$-1.6\,^\circ\text{C/+}$ $3\,^\circ\text{C}$ (2)	$\pm 0.5\,^\circ\text{C}$ (max) (1)
Temperature range	-55 to $125\,^\circ\text{C}$	-30 to $125\,^\circ\text{C}$	-10 to $30\,^\circ\text{C}$	0 to $100\,^\circ\text{C}$	0 to 100°C
Relative inaccuracy[b]	0.48%	0.44%	4.5%	4.6%	1%
Resolution (T_{conv})	$0.063\,^\circ\text{C}$ (6 ms)	$0.033\,^\circ\text{C}$ (200 ms)	$0.2\,^\circ\text{C}$ (30 ms)	$0.1\,^\circ\text{C}$ (100 ms)	$0.25\,^\circ\text{C}$ (12.5 μs)
Resolution FoM	$14.1\,\text{pJ}\,^\circ\text{C}^2$	$1535\,\text{pJ}\,^\circ\text{C}^2$	$140\,\text{pJ}\,^\circ\text{C}^2$	$220\,\text{pJ}\,^\circ\text{C}^2$	$19\,\text{pJ}\,^\circ\text{C}^2$

[a] Excluding the off-chip digital
[b] Relative inaccuracy (%) = 100 · (maximum error/specified temperature range)

(3σ), as shown in Fig. 5.17 (bottom), which is similar to the first prototype. Offset trimming is slightly worse, resulting in an inaccuracy of $\pm 0.5\,^\circ\text{C}$ (3σ). These results show that DTMOSTs, like BJTs, can be effectively trimmed at a single temperature.

While running at a clock frequency of 25 kHz, the sensor requires only 3.6 nJ to achieve a kT/C limited resolution of 63 mK (rms) in a conversion time of 6 ms. This corresponds to a resolution FoM [15] of $14.1\,\text{pJ}\,\text{K}^2$, which is in line with the state-of-the-art BJT-based temperature sensor presented in previous chapter [3]. The performance of the two DTMOST-based sensor prototypes presented in this chapter are summarized in Table 5.2 and compared to that of other state-of-the-art low-voltage designs. It can be seen that the DTMOST-based designs improve the relative accuracy by at least 2×, while the second prototype achieves the best energy efficiency, evidenced by a resolution FoM of $14.1\,\text{pJ}\,\text{K}^2$. Compared to the second BJT-based prototype presented in Chap. 3, this work achieves similar energy efficiency, but at supply voltages as low as 0.85 V. The price to pay will be ~2× less accuracy, which is acceptable in many applications.

5.3 Conclusions

BJT-based temperature sensors, although capable of achieving good energy efficiency and accuracy over wide temperature ranges, are not capable of operating at low supply voltages, which is an important requirement in battery-powered applications and nano-scale CMOS processes. A fully CMOS compatible device, a DTMOST diode, was shown to be a promising alternative to parasitic BJTs as a temperature sensing element. When operated in weak inversion, a DTMOST enables sub-1V operation, while achieving a reasonably high accuracy over a wide temperature range. In this chapter two sensor prototypes based on such sensing elements were demonstrated. The first prototype enabled an apples-to-apples comparison with BJTs, and resulted in the conclusion that DTMOSTs are indeed only a factor 2× less accurate in the chosen 160 nm process. Employing a fully inverter-based 2nd-order zoom-ADC, the second prototype achieved a FoM of 14.1 pJ K^2, which was in line with state-of-the-art BJT-based sensors, back in 2014. When operating at a supply voltage of 0.85 V and after a single temperature trim, the sensor achieved an inaccuracy of $\pm 0.4\,^{\circ}$C (3σ) from -40 to $125\,^{\circ}$C. These results prove that DTMOSTs could be considered as the temperature sensors of choice when sub-1V, high accuracy, and energy efficiency are key requirements.

References

1. M.A.P. Pertijs, K.A.A. Makinwa, J.H. Huijsing, A CMOS temperature sensor with a 3σ inaccuracy of $\pm 0.1^{\circ}$C from -55°C to 125°C. IEEE J. Solid State Circuits **40**(12), 2805–2815 (2005)
2. F. Sebastiano et al., A 1.2-V $10 - \mu$W NPN-based temperature sensor in 65-nm CMOS with an inaccuracy of $\pm 0.2^{\circ}$C (3σ) from -70°C to 125°C. IEEE J. Solid State Circuits **45**(99), 2591–2601 (2010)
3. K. Souri, K.A.A. Makinwa, A 0.12mm^2 7.4μW micropower temperature sensor with an inaccuracy of $\pm 0.2^{\circ}$C (3σ) from -30°C to 125°C. IEEE J. Solid State Circuits **46**(7), 1693–1700 (2011)
4. H. Lakdawala et al., A 1.05V 1.6mW 0.45°C 3σ-resolution $\Delta\Sigma$-based temperature sensor with parasitic-resistance compensation in 32nm digital CMOS process. IEEE J. Solid State Circuits **44**(12), 3621–3630 (2009)
5. K. Ueno, T. Asai, Y. Amemiya, Low-power temperature-to-frequency converter consisting of subthreshold CMOS circuits for integrated smart temperature sensors. Sens. Actuators A Phys. **165**(1), 132–137 (2011)
6. P. Chen et al., A time-domain SAR smart temperature sensor with a 3σ inaccuracy of -0.4°C/$+0.6^{\circ}$C over a 0°C to 90°C range. IEEE J. Solid State Circuits **45**(3), 600–609 (2010)
7. A.J. Annema, Low-power bandgap references featuring DTMOS. IEEE J. Solid State Circuits **34**(72), 949–955 (1999)
8. M. Terauchi, Selectable logarithmic/linear response active pixel sensor cell with reduced fixed-pattern-noise based on dynamic threshold MOS operation. Jpn. J. Appl. Phys. **44**(4B), 2347–2350 (2005)
9. M. Terauchi, Temperature dependence of the subthreshold characteristics of dynamic threshold MOSFETs and its application to an absolute-temperature sensing scheme for low-voltage operation. Jpn. J. Appl. Phys. **46**(7A), 4102–4104 (2007)

10. K. Souri, Y. Chae, Y. Ponomarev, K.A.A. Makinwa, A precision DTMOST-based temperature sensor, in *Proceedings of ESSCIRC*, 2011, pp. 279–282
11. V. Gromov et al., A radiation hard bandgap reference circuit in a standard 0.13μm CMOS technology. IEEE Trans. Nucl. Sci. **54**(6), 2727–2733 (2007)
12. K. Souri, Y. Chae, F. Thus, K.A.A. Makinwa, A 0.85V 600nW all-CMOS temperature sensor with an inaccuracy of $\pm0.4°$C (3σ) from -40 to 125°C, in *Digest of Technical Papers (ISSCC)*, 2014, pp. 222–223
13. B. Razavi, *Design of Analog CMOS Integrated Circuits* (McGraw-Hill, New York, 2001)
14. Y.-H. Lam, W.-H. Ki, CMOS bandgap references with self-biased symmetrically matched current-voltage mirror and extension of sub-1-V design. IEEE Trans. Very Large Scale Integr. VLSI Syst. **18**(6), 857–865 (2010)
15. K.A.A. Makinwa, Smart temperature sensor survey, http://ei.ewi.tudelft.nl/docs/TSensor_survey.xls
16. M. Law, A. Bermak, H. Luong, A sub-μW embedded CMOS temperature sensor for RFID food monitoring application. IEEE J. Solid State Circuits **45**(6), 1246–1255 (2010)
17. Y.S. Lin, D. Sylvester, D. Blaauw, An ultra low power 1V, 220nW temperature sensor for passive wireless applications, in *Proceedings of CICC*, 2008, pp. 507–510
18. C.-K. Wu et al., A 80kS/s 36μW resistor-based temperature sensor using BGR-free SAR ADC with an unevenly-weighted resistor string in 0.18μm CMOS. IEEE J. Solid State Circuits **34**(72), 949–955 (1999)

Chapter 6
Conclusions

In this thesis, the development of energy-efficient, accurate smart temperature sensors for wireless temperature sensing applications has been investigated. It has been shown that the existing temperature sensors prior to the start of this research were ill suited for use in such applications, where energy efficiency and low cost are critical requirements. In the following, first the main findings of this research are discussed. The other applications of the developed techniques are then presented, followed by some proposals for the future improvements.

6.1 Main Findings

The main findings of this thesis are as follows:

- After a single temperature trim, a BJT-based temperature sensor achieved an inaccuracy of $\pm 0.2\,°C$ (3σ) from -55 to $125\,°C$ across different lots of a $0.16\,\mu m$ CMOS process. In older CMOS processes and using a similar sensing approach, slightly better accuracy was achieved, e.g., $\pm 0.1\,°C$ (3σ) in a $0.7\,\mu m$ CMOS process. Down to the $0.16\,\mu m$ node, therefore, CMOS scaling does not significantly impact the accuracy of BJT-based temperature sensors (Chap. 4).
- Of the various CMOS-compatible temperature sensing elements, substrate PNPs are best suited for use in wireless temperature sensing applications. This is because they can be operated at bias currents down to tens of nA with minimal impact on their accuracy. Moreover, they exhibit a well-defined process spread, which can be effectively trimmed at a single temperature (Chap. 1).
- A two-step zoom-ADC was developed, which combines the speed of a coarse SAR-ADC, with the high resolution/accuracy of a fine $\Delta\Sigma$-ADC. In contrast to conventional $\Delta\Sigma$-ADCs, the full-scale range of the fine converter is considerably reduced, which notably relaxes various key requirements such as the number of

© Springer International Publishing AG 2018
K. Souri, K.A.A. Makinwa, *Energy-Efficient Smart Temperature Sensors in CMOS Technology*, Analog Circuits and Signal Processing,
DOI 10.1007/978-3-319-62307-8_6

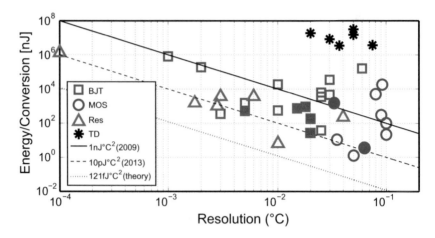

Fig. 6.1 Energy per conversion versus resolution for various smart temperature sensors [1]. The *solid and dashed FoM lines* represent the state of the art in 2009 and 2013, respectively. The *solid symbols* indicate the various sensors realized during this research. The theoretical FoM line of a BJT-based front-end using the (worst-case) bias current scaling ($p = 2$) is illustrated by the *dotted line* (Sect. 2.4.1)

$\Delta\Sigma$-cycles and the DC gain and swing of the loop filter. In this architecture, both conversion time and power efficiency can be improved, resulting in a substantial improvement in the energy efficiency. The fact that dynamic correction techniques can still be used in the fine conversion phase ensures that the accuracy of the zoom-ADC can be as good as that of conventional $\Delta\Sigma$-ADC architectures (Chap. 3).

- BJT-based temperature sensors are also applicable to automotive applications, where high temperature operation ($>150\,°C$) is desired. It was shown that by optimizing the emitter area and bias current of substrate PNPs, the impact of their saturation current I_S at high temperatures can be well mitigated. Furthermore, robust circuit techniques were employed to cope with the various leakage currents at such temperatures. The sensor achieved an accuracy similar to that of state-of-the-art thermal diffusivity (TD) sensors, while drawing more than an order of magnitude less current (Chap. 4).

- DTMOSTs were shown to be a promising alternative to parasitic BJTs for temperature sensing at low supply voltages. When operated in weak inversion, a DTMOST enables sub-1V operation, while achieving a reasonably high accuracy over a wide temperature range. In a $0.16\,\mu$m process, after a single-point trim, an inaccuracy of $\pm 0.4\,°C$ (3σ) was achieved over the military temperature range, which is only a factor $2\times$ worse compared to, similarly trimmed, BJT-based sensors (Chap. 5).

Figure 6.1 plots the dissipated energy per conversion versus sensor resolution of various smart temperature sensors [1]. The different sensor prototypes developed during this research are denoted by the bold symbols in this plot. As shown, the efficiency of smart sensors published prior to the start of this research (in 2009) was limited to about $1\,\text{nJ}\,^\circ\text{C}^2$, denoted by the solid FoM line. Since then, the work described in this thesis (among others) has improved the energy efficiency of BJT-based temperature sensors by a remarkable two orders of magnitude, represented by the dashed FoM line of $10\,\text{pJ}\,^\circ\text{C}^2$. One of the designs in this work, achieved a resolution FoM of $11\,\text{pJ}\,^\circ\text{C}^2$, which defined the state of the art when it was published in 2012 [2]. Furthermore, this research presented the first DTMOST-based smart temperature sensors, which achieved the highest relative accuracy in the class of MOSFET-based temperature sensors, after only a single temperature calibration [3, 4]. To date, the work presented in [4] is the only all-CMOS temperature sensor to achieve both sub-1V operation, and a high accuracy of $\pm 0.4\,^\circ\text{C}$ (3σ) over a wide temperature range (-40 to $125\,^\circ\text{C}$), while employing only a single room-temperature trim. It also achieves good energy efficiency: with a FoM of $14\,\text{pJ}\,^\circ\text{C}^2$.

Last but not least, different flavors of the first BJT-based sensor prototype (Chap. 4) were productized by NXP Semiconductors, namely the PCT2075 [5] and the PCT2202UK [6]. Moreover, the first prototype was also integrated into a multi-sensor platform for smart RFID tag applications, during the EU-funded Pasteur project led by NXP Semiconductors. The tag is used to predict the expiration date of perishable goods, based on the history of conditions (including temperature), in which the items are stored and transported (https://nxp-rfid.com/pasteur-smart-sensor-tag-based-nxps-ucode-i2c-wins-food-valley-award-2013/). Future products based on the 2nd-order zoom-ADC are in the pipeline.

6.2 Other Applications of This Work

The zoom-ADC technique described in this thesis can also be used to improve the energy efficiency of the general-purpose ADCs. However, one drawback of the proposed architecture is that the coarse and fine conversions are performed sequentially. Therefore, the references of the fine $\Delta\Sigma$-ADC are set based on the results of the coarse SAR conversion, and thus are fixed throughout the fine conversion. This, in turn, imposes a limitation on the input signal bandwidth, which must remain within certain limits during the fine conversion step, to avoid out-of-range errors. This property matches the requirements of most instrumentation and sensor readout applications, in which high resolution/accuracy and energy efficiency are key requirements, while the input signal bandwidth is usually low, e.g., less than 100 Hz. To demonstrate this, a general-purpose incremental zoom-ADC was developed [7]. It combines a 6-bit coarse SAR conversion with a 15-bit fine $\Delta\Sigma$-ADC, and employs a fully inverter-based structure. In a conversion time of 40 ms, it achieves a resolution of 20 bits, while drawing $6\,\mu\text{A}$ from a 1.8 V supply. By the use of dynamic error correction techniques, it achieved an offset of $1\,\mu\text{V}$ and an INL of 6 ppm. It also obtained a Schreier FoM of 182.7 dB, which defined the state of the art in the class of incremental ADCs, when it was published in 2013 [7].

In a recent work [8], parallel operation of the coarse and fine conversions in the zoom-ADC is presented, which allows the input signal bandwidth to be increased without causing out-of-range errors. Since the coarse SAR-ADC is continuously running, the references of the fine $\Delta\Sigma$-ADC are dynamically updated, and thus rapidly track the input signal. The resulting zoom-ADC achieves a dynamic range of 107.5 dB in a signal bandwidth of 20 kHz, which meets the requirements of audio codecs. By drawing 0.92 mA from a 1.8 V supply, it achieves a competitive Schreier FoM of 178.3 dB, but occupies significantly smaller area (0.16 mm^2), when compared to state-of-the-art ADCs with similar performance. This is due to the advantages offered by the zoom-ADC architecture, which enables a simple circuit design and a compact silicon realization.

6.3   Future Work

As presented, the BJT-based second prototype in Chap. 4 achieves a FoM of 11 pJ °C^2. As described in Chap. 2, the theoretical energy efficiency limit for the BJT-based sensors is 121 fJ °C^2 (see Fig. 6.1), which is more than one order of magnitude smaller than that of state-of-the-art sensors. This is partially because in the theoretical FoM calculations of Chap. 2, the power dissipated in the front-end's bias circuit and ADC were neglected. After applying the same assumption to our sensor prototype, however, results in a FoM of 3.5 pJ °C^2, which is still much larger than the theoretical limit. The main reason for this large gap is the fact that the sensor's resolution is dominated by the kT/C noise of its switched-capacitor readout. Considering the switched-capacitor 2nd-order zoom-ADC presented in Chap. 4, and assuming that the output noise is dominated by the sampling of ΔV_{BE}, the associated voltage noise during the fine conversion can be approximated by:

$$v_{n,\Delta V_{BE}}^2 \approx \frac{kT}{(n+1) \cdot C_U \cdot N_{\Delta\Sigma}}, \tag{6.1}$$

where k and T represent the Boltzmann constant and temperature in Kelvin, respectively. The parameter n is the integer part of X, as presented in Eq. (3.7), C_U is the DAC's unit sampling capacitor, and $N_{\Delta\Sigma}$ is the number of $\Delta\Sigma$ cycles during the fine conversion step. For the second prototype presented in Chap. 4, n = 14 at room temperature, C_U = 120 fF, and $N_{\Delta\Sigma}$ = 128. By applying these values to Eq. (6.1) and after multiplying by the sensitivity $S_{\Delta V_{BE}}^{D_{out}}(T)$ [see Eq. (2.19)], an output noise of $\approx$16 mK can be calculated, which is in-line with the measurement results presented in Sect. 4.2.4. Therefore, even by neglecting the power dissipation in the bias circuit and the ADC, it is not possible to achieve a FoM better than 2.2 pJ °C^2. Increasing the size of C_U to reduce noise will not improve things, as it will require a proportionally longer settling time, and thus more energy dissipation. This observation suggests that in order to further improve energy efficiency, a continuous-time (CT) readout should be used, thus avoiding the noise aliasing inherent to the use of switched-capacitor circuits. The challenge will then be to tackle the resulting sensitivity to sampling clock jitter, and how to implement the dynamic element matching needed for accuracy.

As discussed in Chap. 4, and can be seen from Eq. (2.9), the lot-to-lot spread observed in the fitting parameters A and B corresponds to a major loss of accuracy, e.g., a shift of 0.5 °C was observed between two lots, at room temperature. Over the full temperature range and across multiple process lots, this error is expected to be even larger. To overcome this, the optimum A and B for each batch should be determined via a batch calibration, i.e., by characterizing a few samples from each batch over the full temperature range. Better accuracy can be obtained by an individual calibration, preferably only at room-temperature. Voltage calibration, as described in [9], can be used to significantly speed up such room-temperature calibration. Nevertheless, characterization over temperature, as required for the batch calibration, will still dictate the use of climate chambers or oil-baths. The associated calibration time, and therefore extra cost, is a major limiting factor on the accuracy of commercial products. Further development of calibration techniques based on electrical measurements in order to eliminate the need for over-temperature characterization will be extremely valuable.

Although voltage calibration is an effective alternative to thermal calibration, it still requires accurate knowledge of the externally applied reference voltage, e.g., a $\pm 200 \mu V$ error corresponds to a calibration error of ± 0.1 °C [9]. Such accurate voltage measurement is not always possible in standard high-volume test equipment, and may call for customized test setups, which is not desirable. Since thermal-diffusivity (TD) sensors are apparently able to achieve high accuracy without calibration [10], an alternative method would be to use them as on-die reference sensors during the calibration process. This would eliminate the need for precision voltage measurements by the tester. The TD sensor's relatively large operating power would then only be dissipated during the calibration step. The challenge, however, will be to design a single readout circuit which can be efficiently shared between the TD and the main sensor, e.g., a BJT- or DTMOST-based sensor, with minimal power and area overhead. It should be noted, however, that the uncalibrated accuracy of TD sensors over different process lots has not yet been demonstrated and requires further investigations.

References

1. K.A.A. Makinwa, Smart temperature sensor survey, http://ei.ewi.tudelft.nl/docs/TSensor_survey.xls
2. K. Souri, Y. Chae, K.A.A. Makinwa, A CMOS temperature sensor with a voltage-calibrated inaccuracy of ±0.15°C (3σ) from −55 to 125°C, in *Digest of Technical Papers (ISSCC)*, 2012, pp. 208–209
3. K. Souri, Y. Chae, Y. Ponomarev, K.A.A. Makinwa, A precision DTMOST-based temperature sensor, in *Proceedings of ESSCIRC*, 2011, pp. 279–282

4. K. Souri, Y. Chae, F. Thus, K.A.A. Makinwa, A 0.85V 600nW all-CMOS temperature sensor with an inaccuracy of $\pm 0.4°$C (3σ) from -40 to $125°$C, in *Digest of Technical Papers (ISSCC)*, 2014, pp. 222–223
5. PCT2075 data sheet, NXP Semiconductors, 2014, www.nxp.com
6. PCT2202UK data sheet, NXP Semiconductors, 2015, www.nxp.com
7. Y. Chae, K. Souri, K.A.A. Makinwa, A 6.3µW 20bit incremental zoom-ADC with 6ppm INL and 1µV offset. IEEE J. Solid-State Circuits **48**(12), 3019–3027 (2013)
8. B. Gonen, F. Sebastiano, R. van Veldhoven, K.A.A. Makinwa, A 1.65mW 0.16mm^2 dynamic zoom-ADC with 107.5dB DR in 20kHz BW, in *Digest of Technical Papers (ISSCC)*, 2016, pp. 282–283
9. M.A.P. Pertijs, J.H. Huijsing, *Precision Temperature Sensors in CMOS Technology* (Springer, Dordrecht, 2006)
10. C.P.L. van Vroonhoven, D. d'Aquino, K.A.A. Makinwa, A thermal-diffusivity-based temperature sensor with an untrimmed inaccuracy of $\pm 0.2°$C (3σ) from $-55°$C to $125°$C, in *Digest of Technical Papers (ISSCC)*, 2010, pp. 314–315

Index

A

Accuracy, xiii–xvi, 5, 6, 9–11, 13–16, 20, 24, 25, 31, 33–35, 38, 46, 47, 52, 56, 57, 60, 61, 63–65, 68, 73, 75, 79, 80, 82, 83, 85, 87, 88, 91, 92, 95, 97, 99–100, 104–107, 109–111, 113

Active radio frequency identification tags, 3–5

ADC. *See* Analog-to-digital converter (ADC)

a-trim, xiv, 56, 69, 79, 87, 95–97, 105

Analog front-end, 23, 37, 59–64, 70, 74, 82–84

Analog-to-digital converter (ADC), xiv–xvi, 9, 13, 15, 16, 19, 23–26, 30, 32–35, 37–54, 57, 59, 60, 62, 64–73, 75, 77, 79–80, 84–85, 87, 93–95, 100, 102–105, 107, 109–112

Auto-zeroing (AZ), 46, 73

B

Bandgap temperature sensors, 23–25

Bandgap voltage, 19, 21, 22, 82

Bandgap voltage reference, 7, 46, 92, 93

Batch-calibrated, 15, 67, 94

Batch-calibration, 9

Batch-to-batch spread, xv, 78–79, 87

Battery-powered, 2, 5, 107

β-compensation, 94

β_F-compensation, 61, 74, 98

β_F-dependent, 61, 98

Bipolar junction transistors (BJT)-based sensor, xiv–xvi, 11, 16, 55, 57, 87, 91–93, 95, 107, 111, 112

Bipolar junction transistors (BJT)-based temperature sensors, 7, 19–35, 57, 88, 93, 106, 107, 109, 110

Bitstream (bs), 33, 43, 45, 47, 65–67, 71, 103

Bottom-plate sampling, 67

bs. *See* Bitstream (bs)

C

Calibration, xiii, xv, xvi, 5, 6, 9, 11, 13–15, 25, 31, 34, 52, 54–57, 75–79, 81, 87, 97, 106, 111, 113

Capacitor digital-to-analog converter, 52, 72, 84, 85, 96

CDS. *See* Correlated double-sampling (CDS)

Charge balancing, xv, 23, 33, 34, 45, 51, 52, 65, 87

Charge injection, 74, 75

Charge-pump, 101, 102

Chopping, 33, 46, 62, 67, 74, 83, 99, 100

Clinical temperature range, 24

Coarse conversion, 42–44, 46, 47, 49, 51, 52, 64–66, 70–74, 103

Complementary-to-absolute-temperature (CTAT), xiii, 7, 21, 22, 92, 93, 104, 105, 113

Compound annual growth rate (CAGR), 3

Continuous-time (CT), xvi, 112

Conversion time, xiv, 5, 6, 14, 15, 26, 27, 32–34, 37, 38, 41, 44, 57, 64, 67, 68, 72, 75, 78, 80, 86, 87, 106, 110, 111

Correlated double-sampling (CDS), 67, 74

© Springer International Publishing AG 2018
K. Souri, K.A.A. Makinwa, *Energy-Efficient Smart Temperature Sensors in CMOS Technology*, Analog Circuits and Signal Processing, DOI 10.1007/978-3-319-62307-8

Cost-effective, 5, 7
CT. *See* Continuous-time (CT)
CTAT. *See* Complementary-to-absolute-
    temperature (CTAT)
Current gain β_F, 7, 22, 61–62
Current-voltage mirror (CVM), 98, 99, 105
Curvature, 55, 57, 60, 67, 75, 104
CVM. *See* Current-voltage mirror (CVM)

D
DAC. *See* Digital-to-analog converter (DAC)
Decimation filter, 45, 67, 75, 85, 94, 99, 104
$\Delta\Sigma$-ADC, xiv, 23, 32–34, 37, 38, 44, 46,
    48–50, 52, 57, 64, 70, 84, 85, 87, 94,
    109–112
$\Delta\Sigma$ converters, xiv, 33, 42, 64, 94
DEM. *See* Dynamic element matching (DEM)
Differential nonlinearity (DNL), 52–54, 80
Digital back-end, xiv, 30, 37, 38, 55, 60, 67,
    68, 75, 85, 94, 100
Digitally assisted, 38, 40
Digitally assisted analog, 38
Digital-to-analog converter (DAC), 44–46,
    52–54, 72, 84, 85, 96, 102, 103, 112
DNL. *See* Differential non-linearity (DNL)
DTMOSTs. *See* Dynamic threshold MOS
    transistors (DTMOSTs)
Dynamic correction techniques, xiv, 14, 57,
    110
Dynamic element matching (DEM), 24, 33, 46,
    52–54, 63–65, 73, 77, 83, 93, 99, 100,
    103, 104, 113
Dynamic error correction techniques, 33, 34,
    111
Dynamic range, xiv, 24, 25, 37, 38, 42, 112
Dynamic threshold MOS transistors
    (DTMOSTs), xiii, xv, xvi, 11–12, 15,
    16, 91–100, 105–107, 110, 111, 113

E
Effective number of bits (ENOB), 25, 34, 41,
    46, 48, 49, 51
Electro-thermal filter (ETF)-based sensors, 9,
    10, 15
Energy efficiency, xiii–xvi, 5, 6, 13–16, 19,
    26–34, 37, 38, 49, 57, 59, 68–70, 72,
    80, 87, 91, 106, 107, 109–112
Energy-efficiency gap, xiv, 15, 31–32, 57
Energy per conversion, 13, 26, 28, 30, 31, 110
ENOB. *See* Effective number of bits (ENOB)
Environmental monitoring, 5, 25
ETF. *See* Electro-thermal filter (ETF)

F
Feed-forward, 70, 73, 102, 103
Figure of merit (FoM), xiii, xvi, 6, 13–15, 26,
    28–34, 37, 38, 68, 69, 80–82, 106, 107,
    110–112
Fine conversion, xiv, 42, 45, 46, 52, 57, 64–67,
    70, 71, 73–75, 79, 80, 87, 100, 103,
    110–112
1st-order zoom-ADC, 16, 49–54, 59, 64, 65,
    68, 69, 95
FoM. *See* Figure of merit (FoM)
Forward current gain β_F, 22, 61–62
Front-end, xiv, 2, 13, 23, 26–33, 37, 54, 59–64,
    70, 73, 74, 82–84, 93, 95, 98–100,
    103–105, 110
Fully differential, 65, 67, 72

G
General-purpose, xvi, 13, 111
Generic bandgap temperature sensor, 23–25,
    38
Guard-band, 70, 72
Guard-band cycle, 65
Guard-banding, 46–50, 102

I
Inaccuracy, xiv, xv, 6, 7, 9, 14, 15, 19, 31,
    33, 55–57, 62, 67–69, 74–76, 78–82,
    85–87, 91, 92, 94–97, 104–107, 109,
    110
Incremental analog-to-digital converter, 111
Incremental mode, 32, 48
INL. *See* Integral nonlinearity (INL)
Integral nonlinearity (INL), 52, 111
Integrator, xv, 33, 44–46, 49–52, 64–67,
    70–74, 84, 102, 103
Internet of things (IoT), 3
Inverter-based, xv, 100, 102–104, 107
Inverter-based OTA, 73, 84, 103

K
kT/C noise, 67, 95, 112

L
Lateral NPN, 20
Leakage current, xv, 20, 81, 83–85, 94
Loop filters, xiv, 45, 46, 48, 49, 51, 52, 57, 70,
    87, 110
Lot-to-lot spread, 113
Low-cost, 1, 6, 15, 16, 34, 79, 80

Low-cost calibration, 76
Low-swing, 70, 71, 87
Low-swing integrators., 46

M
MEMS frequency references, 25
Military temperature range, 14, 15, 23, 24, 42,
      72, 75, 87, 91, 92, 94, 110
MOSFET-based sensors, 10, 80, 92
Multi-bit digital-to-analog converter, 52

N
Noise bandwidth, 27
Nyquist-rate analog-to-digital converter, 32, 33

O
Offset trim, 56, 57, 76, 95, 96
Operating range, xiii, 4, 5
Out-of-ranging, 46, 47, 65

P
Packaging shift, 79
Packaging stress, 7, 20, 57
Parasitic bipolar junction transistors, 8, 92,
      107, 110
Parasitic vertical PNPs, 7
Passive radio frequency identification tags,
      4–6
Phase-domain analog-to-digital converter, 9
Plastic packaging, xv, 78–79, 87
Proportional-to-absolute-temperature (PTAT),
      xiii–xv, 7, 19, 21–23, 37–39, 55–57,
      59–61, 63, 68, 74, 76–79, 82, 87, 92–94,
      98, 100, 103, 104
Pseudo-differential, 72, 102
Pseudo-DNL, 53, 54, 80
PTAT. *See* Proportional-to-absolute-
      temperature (PTAT)

Q
Quantizer, 33, 34, 44, 46, 48, 52, 67, 70, 73,
      103

R
Radio frequency identification (RFID), xiii,
      3–6, 111
Redundancy, 46–48
Reference sensor, xiii, 5, 54, 76

Residue, 44, 48
Resistor-based sensor, 9, 13, 15, 31
Resolution, xiv, xv, 6, 13–15, 24–27, 30–32,
      34, 35, 37, 38, 40–42, 46, 48–50, 57,
      64, 65, 67–70, 73, 75, 77, 78, 80–82,
      86, 87, 97, 106, 109–112
Resolution figure-of-merit (FoM), 13, 14, 26,
      31, 68, 69, 80–82, 106, 111
RFID tags. *See* Radio frequency identification
      (RFID)

S
Sample-and-hold, 52, 64, 72
SAR. *See* Successive Approximation (SAR)
SAR-ADC, xiv, 32, 34, 42, 43, 57, 64, 70, 94,
      109, 112
SC. *See* Switched-capacitor (SC)
2nd-order zoom-ADC, xv, 48, 49, 59, 70, 72,
      75, 84, 87, 105, 107, 111, 112
Self-biased, 74
Self-biasing opamp, 98
Sensing element, xv, 5–12, 14–15, 19, 91–97,
      107, 109
Shot noise, 27
Single-point calibration, 31
Smart temperature sensor, xiii, 1, 2, 13, 19, 26,
      30–32, 53, 60, 94, 110,
SM CVM. *See* Symmetrically matched
      current-voltage mirror (SM CVM)
Substrate PNP, xiii, xv, 20, 22, 55, 59, 61, 62,
      109, 110
Sub-threshold, 10–12, 91, 93
Successive Approximation (SAR), xiv, 32, 44,
      64–66, 71, 94, 96, 97, 100, 103, 104,
      111
Switched-capacitor (SC), xiv, xvi, 64, 65, 67,
      71, 94, 101, 102, 112
Symmetrically matched current-voltage mirror
      (SM CVM), 98, 99, 105
System-level chopping, 74, 100

T
TD. *See* Thermal diffusivity (TD)
Temperature-to-digital conversion, 100
Thermal calibration, xv, 5, 76–78, 87, 113
Thermal diffusivity (TD), xvi, 9, 14, 81, 86,
      88, 110, 112
Thermal equilibrium, 76
Thermal noise, 8, 13, 14, 27, 33
Thermistors, 1, 80, 81
Trimming, xiv, xv, 5, 9, 15, 19, 53–57, 68,
      75–78, 81, 86, 87, 91, 95, 100, 105, 106

Two-step analog-to-digital converter, 42, 52, 79
Two-step zoom-ADC, xiv, 57, 109

U
Unary-weighted capacitor digital-to-analog converter, 52

V
Voltage calibration, xv, xvi, 76–79, 87, 113

W
Wireless sensor networks (WSNs), xiii, 2, 5
Wireless sensors, 34
Wireless temperature sensing, xiii, 1, 14, 15, 19, 109
WSNs. *See* Wireless sensor networks (WSNs)

Z
Zoom-ADC, xiv, xv, xvi, 15, 16, 34, 42–54, 57, 59, 60, 62, 64, 65, 68–72, 75, 84, 87, 94–96, 100, 102, 104, 105, 107, 109–112

Printed in the United States
By Bookmasters